SpringerBriefs in Applied Sciences and Technology

SpringerBriefs present concise summaries of cutting-edge research and practical applications across a wide spectrum of fields. Featuring compact volumes of 50 to 125 pages, the series covers a range of content from professional to academic.

Typical publications can be:

- A timely report of state-of-the art methods
- An introduction to or a manual for the application of mathematical or computer techniques
- A bridge between new research results, as published in journal articles
- A snapshot of a hot or emerging topic
- An in-depth case study
- A presentation of core concepts that students must understand in order to make independent contributions

SpringerBriefs are characterized by fast, global electronic dissemination, standard publishing contracts, standardized manuscript preparation and formatting guidelines, and expedited production schedules.

On the one hand, **SpringerBriefs in Applied Sciences and Technology** are devoted to the publication of fundamentals and applications within the different classical engineering disciplines as well as in interdisciplinary fields that recently emerged between these areas. On the other hand, as the boundary separating fundamental research and applied technology is more and more dissolving, this series is particularly open to trans-disciplinary topics between fundamental science and engineering.

Indexed by EI-Compendex, SCOPUS and Springerlink.

More information about this series at http://www.springer.com/series/8884

Mariela Agotegaray

Silica-Based Nanotechnology for Bone Disease Treatment

 Springer

Mariela Agotegaray
Instituto de Química del Sur (INQUISUR)
CONICET
Universidad Nacional del Sur
Bahía Blanca, Buenos Aires, Argentina

ISSN 2191-530X   ISSN 2191-5318   (electronic)
SpringerBriefs in Applied Sciences and Technology
ISBN 978-3-030-64129-0   ISBN 978-3-030-64130-6   (eBook)
https://doi.org/10.1007/978-3-030-64130-6

This Springer imprint is published by the registered company Springer Nature Switzerland AG
The registered company address is: Gewerbestrasse 11, 6330 Cham, Switzerland

Contents

Chapter 1
Nanotechnology Based on Silica for the Treatment of Bone Disease

M. Agotegaray, *Silica-Based Nanotechnology for Bone Disease Treatment*,
SpringerBriefs in Applied Sciences and Technology,
https://doi.org/10.1007/978-3-030-64130-6_1

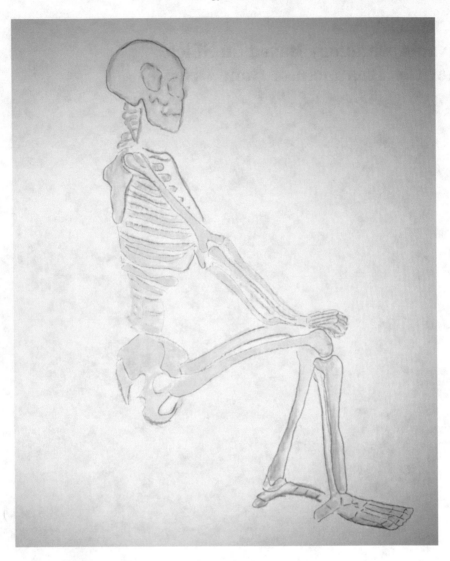

Mariela Agotegaray
   Illustrations by Verónica Florin and Mariela Agotegaray

Mariela Agotegaray is Biochemistry and Doctor in Chemistry. Professor at National University of the South and Scientific Researcher at CONICET, Argentina

Verónica Florin is Director in Visual Arts, Bahía Blanca, Argentina

To Uma, who inspired this book.
To my grandad who showed me
that I need to look forward my bones
and join them to enjoy my wonderful life.
Thanks to them.

When we talk about bones we usually think about the cold and scary image of a skeleton, which most of us were afraid of as little ones. On the other hand, in that mannequin standing in front of the anatomy class that only inspired us enormous anxiety to think that we should learn all and each one of the names of the 206 bones that make it up. However, it is not like that.

The skeleton and each of the bones that make it up have not only a structural function, but also have finely regulated physiological functions that contribute in an extraordinarily active way to the functioning of the entire organism.

Delving into the functions of the bone, we can go far beyond anatomy, histology and physiology. We can delve into the ancestral meaning of bone and from there we will understand many scientific aspects of its functionality.

"The indestructible force of life". This is how Clarissa Pinkola Estés defines bones in her book "Women who run with wolves".

In archetypal symbology bones represent strength. They can create themselves and regenerate constantly. They are structure, but at the same time inside, they have the flexibility to house life. Moreover, when they leave life they cannot be destroyed; it is difficult to burn them and it is almost impossible to spray them. This is how bones resist, remain, and persist. They are the material trace of the passage through this earthly life of many living beings, of all human beings.

It is not common to find metaphors in science. Metaphors generally belong to the magical world of poetry, the art of literature … but in this book, I allow myself to metaphorize from art to science. Because I think, doing science is making art. It is to create with one purpose: explore and discover to understand and help. It is now that I begin to understand why at some point in my life, without knowing it, I chose the path of science to develop my working life. Because until a while ago I did not understand. Now I do, and it has to do with the fact that we develop art to serve humanity through science. To help in all aspects to grow and have a better quality of life. Because by doing science we understand the functioning of life, we delve into the fundamental mechanisms that are the pillars that sustain us. From there, with the development of knowledge and applying what we learn we push ourselves. Therefore, we grow, we improve and it is from science that we can transcend and enter unknown worlds.

That is why I dare in this book to metaphorize science with art. Because science is an art in itself. It is there where the symbology of bone transports me to the pillar of our lives. The skeleton is made up of bones. Together, united and in harmony they give solidity to the body, they allow support and movement; they make a cavity to house the life of organs and tissues. They have evolved from ancient times to today and keep doing it and they will continue to do so silently. Metaphorizing with the archetypal symbology the assembly of each bone is related to the search and union of each of the most solid aspects of our lives.

The spirit of this book is related to addressing from the scientific point of view (one of the aspects that are part of my life and that I enjoy every day) the characteristics and functionality of the bone. In addition, the different diseases that afflict it and the most recent scientific advances of the last five years related to amorphous silica-based materials aimed at regeneration and treatment. The figures have been illustrated and painted by hand to deconstruct and facilitate reading.

I thank to the publisher for the opportunity to capture in this writing the strong interwoven that exists in my life with science and art. In this case, the bone tissue. The indestructible force of life. I also thank to Verónica for taking on the challenge of drawing for this book.

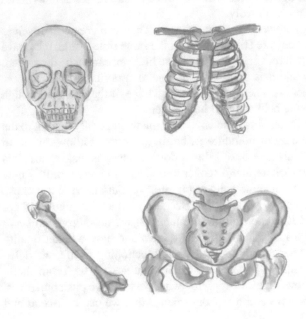

Reference

C. P. Estés, *Women Who Run with the Wolves: Contacting the Power of the Wild Woman* (Random House, 2008)

Chapter 2
Anatomy, Histology and Physiology of Bone

The human skeleton is made up of a set of bones linked together. It is an osseocartilaginous structure, composed of bone and cartilage, which is formed in the fetal period. It is then replaced by what is called a replacement bone while the cartilaginous tissue persists only at certain sites such as in joints, in the nasal septum, in the intercostal area, among others.

In this chapter, the fundamental aspects related to the anatomical and physiological characteristics of the bone will be described, essential to then be able to understand and develop nanomaterials suitable for the treatment of the various pathologies that affect the bone system.

2.1 General Aspects of the Anatomy of the Bone System

Bones are hard and resistant structures or pieces that, in addition to providing support to the muscles, can have two functions:

1. Protective: a set of bones connects to form cavities that house organs. For example, the skull or eye orbits.
2. Articular: that give rise to the mobile joints that join active bones in motor functions. They are joint-by-joint capsules, ligaments and muscles.

The human skeleton has 206 bones in the adult, without taking into account the sutural bones of the skull and the sesamoids, which are small and rounded bones that are embedded in tendons subjected to tension and compression forces.

For a better description, the skeleton can be divided into two portions as **axial** skeleton and **appendicular** skeleton (Fig. 2.1).

The axial skeleton is made up of the skull, the spine and the thorax. The skull and spine house the central nervous system, with its spinal nerves. On both sides of

M. Agotegaray, *Silica-Based Nanotechnology for Bone Disease Treatment*,
SpringerBriefs in Applied Sciences and Technology,
https://doi.org/10.1007/978-3-030-64130-6_2

Fig. 2.1 Skeleton

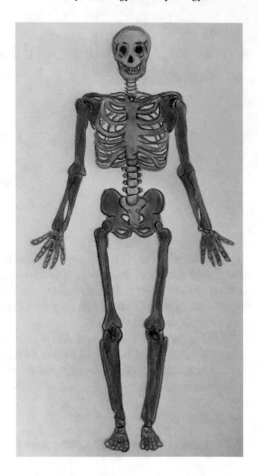

the middle region of the spine the ribs emerge, which articulate in front (except the floating ones) with the sternum. Together, the spine, the sternum ribs and the cartilage involved form the thorax, a cavity that houses and protects important organs of the respiratory, circulatory and digestive system.

The appendicular skeleton is made up of the upper and lower limbs. The pectoral waist formed by the scapula and the clavicle joins the upper limbs to the thorax while the pelvic girdle, constituted by the coxal bones, is located at the lower end of the spine and allows the connection of the lower limbs. The coxal bones, sacrum and coccyx give rise to the so-called pelvic cavity, which houses organs of the digestive, reproductive and excretory system.

The following details some of the most important bones that form the skeleton and its organization to a better comprehension related to location and functions.

Parts of the skeleton and most important bones		
AXIAL	Cranium	Frontal, ethmoid, sphenoid, parietal, temporal, occipital, maxilla, mandible, lacrimal, nasal, zygomatic, vomer, palatine.
	Spine	Cervical, thoracic, lumbar, sacrum, coccyx.
	Thorax	Sternum, ribs.
APENDICULAR	Upper limbs	Scapula, clavicle, humerus, radius, ulna, carpus, metacarpus, phalanges.
	Lower limbs	Coxal, femur, patella, tibia, fibula, tarsals, metatarsals, phalanges.

2.1.1 How Are the Bones from Outside?

Observing the bones from the outside, we can find particularities in terms of their shape and their surface.

Bones can acquire different forms that relate to their location and function. Thus, long bones are those in which predominates length over width and thickness. The body is called diaphysis and the extremes arc named as epiphysis. Those bones of the first two segments of the upper and lower limbs are classified as long bones (Fig. 2.2).

On the other hand, short bones have similar measurement in their three dimensions. They present small volume and diverse shapes. Such is the case of the carpal bones and the tarsus (Fig. 2.3).

The flat bones form the body cavities: cranial, orbital, nasal and pelvic. Length and width predominate over its thickness. In addition, they provide ample surfaces for the insertion of various muscles, such as the scapula, the occipital bone and the coxal (Fig. 2.4).

In addition, there are pneumatic bones in the body that create cavities full of air, like some bones in the face. These cavities are called cells when their size is small or breasts when they acquire larger dimensions. And you can find the so-called sesamoid bones, of a very small size comparable to that of a sesame seed in general, attached to a joint (Fig. 2.5).

Fig. 2.2 Long bone

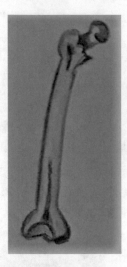

Fig. 2.3 Short bones of the hand

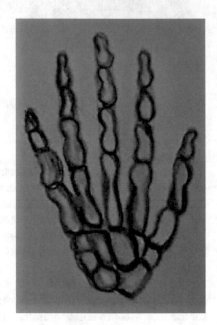

2.1.2 The Bones from Inside

When we talk about bone, sometimes we only think about its structural and supporting function. But internally the bone has essential characteristics for the development of life. And its histology has a lot to do with this.

From the inside, when making a bone cut, two types of structures can be seen: the compact bone that is located in the peripheral area and has the function of protecting

Fig. 2.4 Flat bones of the skull

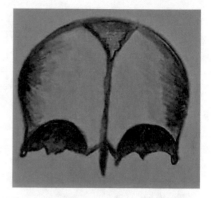

Fig. 2.5 Bones of the knee joint

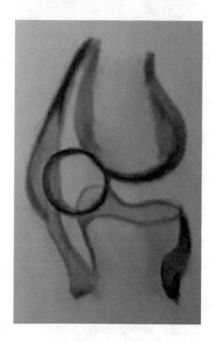

the cancellous bone that is inside. The cancellous bone is made up of trabeculae that communicate with each other, creating spaces where the bone marrow is housed. The function of the trabeculae is to enable greater resistance to the pressure and traction to which the bone is exposed, due to its orientation and using the minimum amount of tissue necessary. Figure 2.6 shows a diagram of a long bone with each of the internal structures. In this type of bone they are better appreciated due to their larger size.

The bone marrow plays a crucial role in the formation and renewal of blood cells, in a process called haematopoiesis. It is found in the medullary cavity in the long bones and in the cavities of the cancellous bone. It is considered a specific organ in the body.

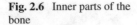

Fig. 2.6 Inner parts of the bone

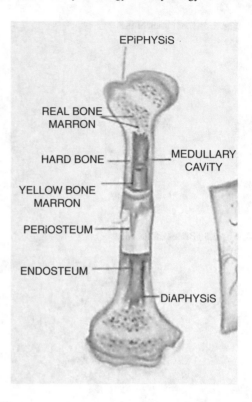

For the focus of this book, related to nanotechnology and amorphous silica based treatments for bone pathologies, it is important to mention the periosteum. It is the membrane that covers the external surface of the bones, except for the places surrounded by articular cartilage and the sites of insertion of ligaments and tendons. Its function is very active in the growth and vascularization processes.

2.1.3   Vascularization and Innervation

Another important part of the anatomy of bones has to do with veins, arteries and nerves. The bone has an interesting vascular richness. This is denoted by the numerous superficial holes called nutrient foramina that deepen inward. They are classified into three orders according to their dimensions.

The arteries are the vascular channels that nourish the bones, just as it happens with other organs and systems of the body. They take different positions according to the part of the bone that is responsible for irrigating. Thus, in all bones there are periosteal arteries that give rise to an abundant arterial network that surrounds each bone. These arteries enter through the smallest nutritional foramina into the nutrient channels where they become vascular. Each bone, in turn, has a main artery that enters through the largest nutritional foramen. This artery is distributed throughout the tissue belonging to that bone that supplies and also nourishes the bone marrow. Each artery has a path that is distributed by a system of lamellae and Havers ducts.

One or two veins accompany each nutrient artery from the deep interior of the bone. It is a return circulation network without valves that originates in cavity centres that generally refer to venous sinuses or dilated sinusoids, after branching and ordering until they reach the periosteum. They have a winding path or can be arranged as highly branched fine ducts depending on the shape of each particular bone.

The nerves reach the bone along with the main nutrient artery. It is a sensory innervation responsible for the sensation of bone pain. These fibers come from the muscle innervation and form a plexus in the periosteum that branches into the terminal glomeruli. They come from the cranial and spinal nerves. Inside the bone they do not necessarily accompany the vascular branching.

All the anatomy related to the vascularization and innervation of the bone system is of fundamental importance in the physiological processes of growth and ossification as well as in fractures, in inflammatory and tumor processes. For this reason, understanding of vascular and nervous anatomy are very important when designing treatments and materials that contribute to bone therapeutics.

2.2 Aspects of Bone Histology, Physiology and Biochemistry

The bones, although they seem to be rigid structures from the functional point of view, are living structures that grow, renew and suffer different impacts depending on the activity to which they are subjected. They can fracture and also suffer from different diseases, such as cancer.

To function properly, bones need a balanced diet that provides the nutrients necessary for the ossification process. On the other hand, this dynamic activity that they maintain is finely regulated by hormones secreted by the pituitary, the thyroid and the genital glands.

A wonderful and unique feature of bone is that it can be repaired and rebuilt after fracture, destruction, or removal by a process called local osteogenesis.

Next, the histological and physiological aspects that allow a fine structural and biochemical synchronization to make the bone become part of something bigger, of an organ system such as the skeleton, will be described.

This part will delve into some structural issues that we mentioned above in order to get to the detailed description of the microscopic aspects of the bone structure, and then delve into the description of functional dynamics.

Bone is a form of connective tissue that is mineralized, adopting the forms of compact or spongy bone formed by trabeculae.

The compact bone, when viewed under a microscope, has ducts that run parallel to each other, and have a diameter of between 10 and 350 μm. These ducts are called Havers' ducts. In the larger ones, the presence of blood vessels, lymphatics

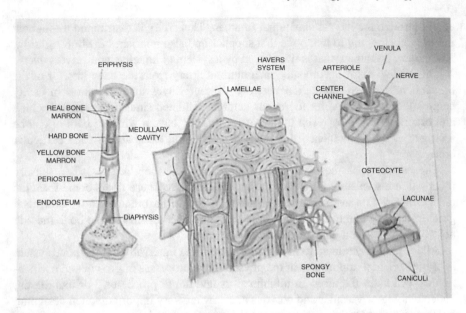

Fig. 2.7 Bone from inside

and nerve fibers can be seen in a longitudinal section, while in the smallest, only one capillary and one venule are observed. Bone lamellae called osteons are arranged concentrically to each Haver canal. Havers ducts communicate with the medulla through transverse ducts named as Volkmann's ducts.

The trabecular bone does not have Havers ducts but has irregular cavities where the bone marrow is located and its lamellae are located around them.

Like other organs, the bone has a coating on the outside and one on the inside. The periosteum covers the bone and is made up of collagen fibers and fibroblasts in its outermost part, while in its deep layer are osteoblast precursor cells, which are a type of bone cell that will be described in more detail later. The inner lining of the bone is called the endostium and is consolidated by precursor cells of osteoblasts and also of osteoclasts, covering the Havers, Volkmann ducts, the medullary canal and all the cavities in the case of trabecular bone.

In this entire bone environment, there are then bone cells, organic matrix and the mineral substance (Fig. 2.7).

2.2.1   Which Are the Cells that Make Up the Bone?

Five types of cells that give life to it and that perform all the necessary functions for its development and maintenance can be fundamentally distinguished in bone. They are osteoprogenitor cells, osteoblasts, osteocytes, bone lining cells, and osteoclasts. In the connective tissue that forms the stroma of the bone, other types of cells such as adipocytes, endothelial cells, and cells of the immune system such as macrophages and mast cells can also be found.

Osteoprogenitor cells are those that give rise to osteoblasts and other cells such as fibroblasts, chondrocytes, adipocytes, and muscles. There are two genes that regulate the differentiation of progenitor cells into osteoblasts: one is the core binding factor A1-CBF A1—that regulates the expression of genes of specific proteins such as osteocalcin, osteopontin and type I collagen, as well as the ligand of the receptor-activator of nuclear factor κB (L-RANK); the other is Ihh (initials of Indian hedgehog) whose expression is necessary for the embryonic development of bone tissue and for the activity of osteoblasts. These osteoprogenitor cells are located on the inner and external surfaces of the bone and in its microvasculature, in the endosteum and in the periosteum, in the deepest layers of each one.

Another type of progenitor cells are those that give rise to osteoclasts. In this way, it is observed that osteoblasts and osteoclasts come from different lines. Thus, the origins of osteoclasts are hematopoietic: cells called "granulocyte-macrophage colony-forming units" GM-CFU. These cells are found in the bone marrow, or they can get through the blood. These cells are activated to differentiate from interaction with cytokines such as interleukins 1, 6, 11; tumor necrosis factor α (TNF-α), interferon and also express on their surface the RANK receptor to respond to L-RANK secreted by osteoblasts.

The **mature osteoblasts** are cuboid in shape and measure between 20 and 30 μm. They are cells that are dedicated to synthesizing a large amount of proteins, for such reasons the nucleus has several nucleoli and the cytoplasm, after staining, can be seen as basophilic, bluish due to its high content of RNA. They are located on the surface in areas of bone tissue formation in the form of a palisade. This location and polarization allows adhesion between neighbouring osteoblasts using cadherins and pinpoint connections. This communication enables the sharing of messengers such as calcium, cytokines and prostaglandins, which results in a coordinated activity for the synthesis and secretion of the organic matrix.

Those osteoblasts that are immersed in the bone matrix are called **osteocytes**. They are responsible for maintaining the matrix and have the particular function of responding to mechanical stimuli. This induces stimulation of gene expression and apoptosis as needed. In this way, they are capable of secreting matrix or degrading it, greatly contributing to the process of calcium homeostasis. The location of osteocytes in the bone structure is very particular: it adapts a starry shape by locating itself in spaces called gaps or osteoplasms. The stellate form is acquired as they deploy cytoplasmic extensions through the canaliculi of the bone matrix (also called calcophore ducts) to establish connections with other neighbouring osteocytes through cleft-type junctions. Osteocytes are smaller than osteoblasts but present different morphologies according to their possible functional states.

Osteoclasts are responsible for bone resorption. Of a large size of up to 100 μm in diameter and reaching up to ten nuclei when they are very active, they present an active brush border where all the activity is concentrated. They have a large number of lysosomes rich in the enzyme tartrate-resistant acid phosphatase (TRAP), employed as a biochemical parameter indicating osteoclastic activity. They also express on the brush border membrane a sodium/bicarbonate pump, an ATP-dependent proton pump, type II carbonic anhydrase and cathepsin K, whose function will be explained later in this chapter.

2.2.2   The Osteoblasts Functions

The main function of osteoblasts is related to the synthesis, organization and mineralization of the bone matrix. Other important functions are related to the control of bone repair caused by a fracture and the regulation of osteoclast activity, through which they participate actively and indirectly in the bone resorption process.

All these functions are finely regulated by molecules of the cytokine type, such as interleukins and growth factors produced by osteoblasts and by the osteocytes themselves.

Produced in the liver, insulin-like growth factor-1 (IGF-1) plays a role as a mediator of growth hormone (GH) and acts at the level of bone formation and remodelling. At the bone level itself, osteoblasts express on their surface the nuclear factor activator receptor $\kappa\beta$ (NF-$\kappa\beta$) ligand which, as will be described below, promotes an action on osteoclasts.

2.2.3   The Bone Matrix

The bone matrix is made up of an organic and a mineral portion. The organic matrix is mostly made up of type I collagen (90%). Type I collagen has certain structural properties that not only contribute to the peculiar structure of bone tissue, but also contribute to bone physiology. It is a non-soluble protein. It is one of the few proteins that contains the amino- acids hydroxyproline and hydroxylysine. In type I collagen these amino acids are arranged in a glycine-X-Y sequence, with X and Y being said amino acids respectively. The RGD sequence that it contains in its structure (Arginine-Glycine-Aspartic Acid) has a physiological function since it is recognized by the superficial integrins of cells, activating mechanisms of action on the extracellular matrix. The –COOH end of one type I collagen molecule interacts with the –NH$_2$ end of the other by overlapping. In turn there are lateral couplings through which induce a packing that forms the characteristic fibrils of the bone. The entire structure is stabilized by hydrogen bonds and pyridinoline bridges. These are formed by the action of the enzyme lysyl oxidase that generates aldehydes in the amino acids lysine and hydroxylysine. These interact with neighbouring hydroxylysines forming the pyridinoline bridges.

The rest of the proteins that make up the matrix belong to the glycoprotein family and are described in the following table.

PROTEIN	FUNCTION
Alkaline phosphatase (ALP)	At an optimal pH of 8.6 it releases inorganic phosphate from phosphoric esters, stimulating the mineralization process. It is produced by osteoblasts.
Glycoproteins with RGD sequence	Osteopontin, fibronectin, thrombospondin, and sialoproteins. Osteoblasts and osteoclasts recognize their common Arg-Gly-Asp sequence and thus represent a key cellular anchoring factor to the matrix and allow migration.
Proteoglycans	They are synthesized by osteoblasts. Chondroitin sulfate (CS), hyaluronan (H), decorin (D), and biglucan (B) can be found. Their functions are aimed at reserving space for mature bone (CS), bone morphogenesis (H) and modulation of growth factors (D, B)
Y-carboxyglutamic acid proteins	These proteins can accommodate $Ca2+$ between carbonyl groups of Y-carboxyglutamic acid. Osteocalcin that is produced by ostoblasts and osteoid protein with carboxyglutamic acid belong to this group. This last one inhibits collagen mineralization in non-bone tissues.

Other proteins such as albumin, haemoglobin, immunoglobulins, and Apo-A1 lipoprotein from plasma are retained in the bone matrix on the surface of the bone mineral. Its concentration exceeds that of plasma, although its physiological role is still unknown.

The mineral fraction of the bone matrix is mainly formed by hydroxyapatite, whose nanometric size and asymmetry generate a large surface area per unit of weight and allow the deposition of ions and water.

2.2.4 Mineralization

The matrix mineralization process is directed by osteoblasts. It is based on the regulation of the quantity of ions, of proteins and on the modulation of said process.

To begin with, osteoblasts generate osteoid vesicles of around 200 nm. Its membrane is enriched with ALP and phospholipids with the ability to interact with calcium ions, from the internal side. The function of ALP is to hydrolyze phosphate esters such as ATP, ADP, AMP and pyrophosphate. Thus the orthophosphate groups it releases interact with calcium to begin to form the mineral calcium phosphate.

When it accumulates inside the vesicles, there comes a time when they rupture and release the amorphous calcium phosphate mineral that is deposited on type I collagen in the extracellular matrix, without too much affinity. Proper pH and increasing concentrations promote hydroxyapatite crystal formation. Within the vesicles there are proteins such as osteocalcin, osteopontin and sialoproteins. Regulation of this process occurs negatively from the high molecular weight proteoglycans in the matrix that inhibit mineral deposition. In turn, the P–O–P bonds of ATP and pyrophosphate bind to calcium phosphate crystals, delaying their maturation to hydroxyapatite. This fine and regulated coordination balances the mineralization process.

2.2.5 Osteocytes

Those osteoblasts that are immersed in the bone matrix are then later called osteocytes. Their activity is related to maintaining the matrix and have the particular function of responding to mechanical stimuli. These induce stimulation of gene expression and apoptosis as needed. In this way, they are capable of secreting matrix or of degrading it, greatly contributing to the process of calcium homeostasis. The location of osteocytes in the bone structure is very particular: it adapts a starry shape by locating itself in spaces called gaps or osteoplasms. The stellate shape is acquired as they deploy cytoplasmic extensions through the canaliculi of the bone matrix (also called calcophore ducts) to establish connections with other neighboring osteocytes through cleft-type junctions. Osteocytes are smaller than osteoblasts and present different morphologies according to their possible functional states: latent, formative or springy. Resorptive function is associated with maintaining balanced levels of calcium in the plasma (Fig. 2.8).

2.2.6 Osteoclast Resorptive Activity

As mentioned above, osteoclasts are large, multinucleated cells that govern the bone resorption process. They are characterized in that they have a brush or scalloped edge to increase the available surface area for the release of enzymes necessary for

Fig. 2.8 Location of different types of cells in bone

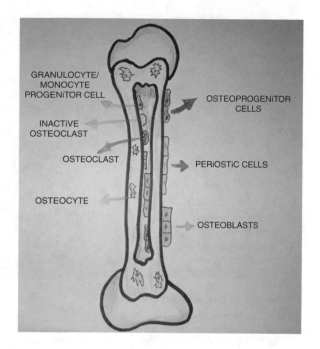

matrix removal. In addition, to accommodate a greater number of ATP-dependent H^+ pumps and in turn to the endocytosis of remaining products of the reabsorption process. To carry out their function, osteoclasts coordinate many functions.

One of the most important has to do with H^+ secretion to decalcify the matrix and dissolve the mineral. In their cytoplasm there is carbonic anhydrase II, responsible for the production of carbonic acid (H_2CO_3) from carbon dioxide and water. The ATP dependent proton pumps located on the brush edge are responsible for expelling the H^+. These channels in turn are coupled to chlorine channels to maintain electroneutrality. Excess bicarbonate from the cytoplasm is eliminated by chlorine ion exchange proteins located in the basolateral membrane, through passive transport. They contain in their cytoplasm numerous lysosomal vesicles filled with hydrolases responsible for matrix degradation, among which are cathepsin K and metalloproteinases (Fig. 2.9).

2.2.7 Bone Remodelling

The bone remodelling process is physiological and occurs at specific sites through a fine regulation of the osteoblasts on the osteoclast activity.

Remodelling begins in situ from a cluster of hematopoietic osteoclast precursor cells, which differentiate into osteoclasts from the effect of cytokines such as TNF-α, IL-1, IL-6, and RANK-L.

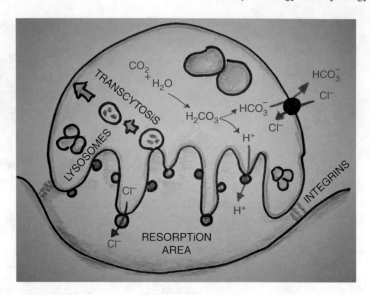

Fig. 2.9 Osteoclastic activity

Once activated, these osteoclasts are anchored to the matrix by the integrins and an area restricted by actin filaments is generated that delimits their site of resorption. Thus the disintegration of the matrix occurs first on the mineral. This is achieved by acidification to pH 4.4 from the release of HCl by the osteoclasts. This occurs through an H^+ ATPase located on the brush edge and a Cl^-/HCO_3^- channel on the inside of the brush edge membrane. It also contributes a chlorine channel coupled to H^+/ATPase. Once the mineral substance is dissolved, the resorption of the organic substance facilitated by cathepsin K, which is released by the osteoclasts themselves and by collagenase that is supplied by the osteoblasts. The products of resorption are removed by osteoclasts through transcytosis. An also mononuclear phagocytes help to this last part of the process by completing elimination.

Bone resorption acquires different morphology according to whether it occurs in the cortical or trabecular area of the bone. In cortical bone, cylindrical channels of about 150 nm in diameter and a few mm in length are formed. While in the trabecular bone, the area acquires a cubic shape and is called Howship's lagoon.

Once resorption is complete, the area must be recomposed. Thus, osteoclasts communicate by cytokines with preosteoblasts and they begin to regroup around the reabsorbed area. After several days these cells begin to produce a cementitious substance that serves as an anchor for the new tissue to come. Preosteoblasts mature into osteoblasts and they begin to produce organic matrix. Then its mineralization occurs. This described process occurs naturally throughout life.

2.2.8 Regulation

The regulation of the resorption process is very fine. The RANK/L-RANK system is a fundamental pillar for the physiological regulation of bone function and is governed by the family of cells of the preosteoblast and osteoblast type. These cells produce L-RANK which is the ligand of the nuclear factor activator receptor κβ. This interaction activates a cascade of downstream transcription factors, such as tumor necrosis factor (TNF), receptor associated factor 6 (TRAF6), nuclear factor-κB (NF-κB), and activator protein 1 (AP-1) in order to regulate cellular homeostasis and differentiation. This system induce the differentiation of preosteoclasts into osteoclasts and thus activate them. Also, preosteoblasts and osteoblasts produce a protein called osteoprotegerin. It is a molecule that circulates and blocks L-RANK when combined with it. In this way, preosteoblasts, osteoblasts and osteocytes govern osteoclastic activity related to bone resorption. In Fig. 2.10 it is schematically represented the operation of these mechanisms.

Recently, other pathway of regulation related to this system has been revealed. Maturing osteoclasts secrets vesicular RANK. This binds osteoblastic RANKL regulating the activity of osteoblasts and thus promoting bone formation. This pathway is named as RANKL reverse signalling. Vesicular RANK interaction with RANKL

Fig. 2.10 Regulation of osteoclast differentiation

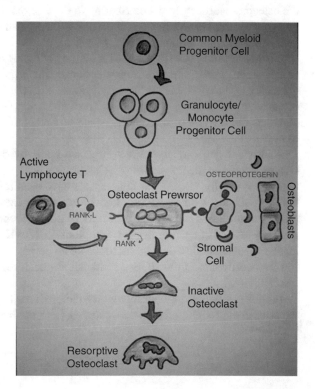

activates Runt-related transcription factor 2 (Runx2) inducing osteoblast to go on positively regulating osteoclastogenesis.

Going deeper into the regulatory mechanisms, it is important to mention that there are not only those mentioned that work locally. Hormonal regulation is also an extremely important factor in bone activity. In this sense, the hypothalamus, the pituitary, the thyroid gland, the parathyroid, the pancreas, the adrenal glands and the gonads are actively involved in the remodelling process.

At the ventromedial nucleus, the hypothalamus has neurons that express leptin receptors. It is a hormone secreted by adipocytes that regulates appetite from these hypothalamic receptors. From these receptors, it also activates, through the parasympathetic nervous system, the inhibition of bone formation through β2-adrenergic receptors located in osteoblasts. In this way the antiosteogenic effect of this hormone is observed.

Thyroid hormones exert their regulation both at the level of osteogenesis and bone resorption. Osteoblasts have T3 receptors that favour the expression of IGF-1 related to the stimulation of osteoid matrix production. While on the other hand, and according to other conditions present in the bone medium, it stimulates the number and activity of osteoclasts.

Parathormone stimulates bone resorption through receptors present in osteoblasts that stimulate the L-RANK pathway. On the other hand, its pulses act at the level of the osteoprotegerin system, contributing to the fine regulation of the process.

For its part, calcitonin acts directly on osteoclastic precursor cell receptors, inhibiting the resorption process.

Growth hormone (GH) regulates the activity of bone through various routes, direct and indirectly. GH induces the production in the liver of IGF-1 that favours the differentiation to osteoblasts. Directly, it stimulates the synthesis of type I collagen, osteocalcin and alkaline phosphatase. Through its action on osteoblasts, it indirectly regulates the activity of osteoclasts, since GH receptors have not been revealed in osteoclasts.

Androgens generate an indirect effect on bone, by mediating the production of GH, especially at the puberty stage. Osteoblasts have androgen receptors that stimulate their activity. Oestrogens deficiencies have been found to induce a decrease in bone mass. This is due to strictly known functions. Oestrogens inhibit osteoclastic precursor formation and also inhibit osteoclastic precursor differentiation. They stimulate the release of osteoprotegerin and block L-RANK. Progesterone, on the other hand, has an anabolic effect on osteoblastic receptors.

Glucocorticoids inhibit IGF-1 synthesis at the level of osteoblasts, decreasing their activity.

Insulin is a very important regulatory factor since it stimulates the deposition of the osteoid matrix and favours the synthesis from the liver of IGF-1.

References

Y. Ikebuchi, S. Aoki, M. Honma, M. Hayashi, Y. Sugamori, M. Khan, Y. Kariya, G. Kato, Y. Tabata, J.M. Penninger, N. Udagawa, Coupling of bone resorption and formation by RANKL reverse signalling. Nature **561**(7722), 195–200 (2018)

M. Latarjet, A.R. Liard, Anatomía humana. Ed. Médica Panamericana (2004)

D.J. Papachristou, E.K. Basdra, A.G. Papavassiliou, Bone metastases: molecular mechanisms and novel therapeutic interventions. Med. Res. Rev. **32**(3), 611–636 (2012)

M.H. Ross, W. Pawlina Histología: Texto y Atlas. Ed. Médica Panamericana (2007)

F. Tresguerres, J. Angel, FISIOLOGIA humana.[director] JAF Tresguerres;[directores asociados] E. Aguilar Benítez de Lugo…[et al.] (2003)

Chapter 3
Human Bone Diseases and Conventional Treatments

Many pathologies affect the bone system. In this chapter the most common ones will be described, which are osteoporosis, cancer and infections. The treatments currently used and their limitations will be discussed.

3.1 Osteoporosis

Osteoporosis is the most common pathology affecting bone. As its name implies, it means "porous bones" and this characteristic of the bone architecture itself is due to a decrease in bone mass, leading to its extreme weakness and fracture. The most affected bones by this pathology are the hip, spine, wrist and long bones (Fig. 3.1).

Fractures are the consequence of this pathology, being approximately in 9 million per year, worldwide. Due to this prevalence, osteoporosis represents a serous public health concern. The risk of morbidity and mortality is intensified in people who suffer bone fractures, not because of the fracture itself but because of the risk of complications associated with hospitalization and immobilization. These include pneumonia, pulmonary thrombosis, and stroke.

The pathophysiology of this disorder is related to disturbances of the balance between the osteoclast-governed resorption process and osteoblast-mediated new bone formation.

Three types of osteoporosis can be distinguished, depending on their ethology and the stage of life in which it manifests.

Type I osteoporosis which represents the highest incidence, occurs in post-menopausal women has to do with the decreased regulation of bone activity by oestrogens, in special estradiol. This hormone plays a crucial role in decreasing the production of cytokines such as IL-1 and TNF that stimulate osteoclast activity. Furthermore, they also inhibit the secretion of M-CSF and IL-6, both cytokines involved in the

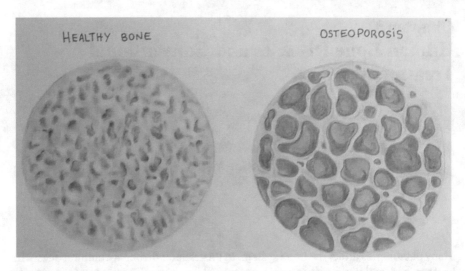

Fig. 3.1 Osteoporosis

differentiation of osteoclast precursors. Thus, during the postmenopausal period, the decrease in estradiol characteristic of this hormonal stage in women, deregulates the resorptive activity and the osteoclasts are much more active, leading to greater resorptive activity.

Type II osteoporosis presents the same pathophysiology but occurs lately, near the seventh or eighth decade of life.

Meanwhile, **secondary osteoporosis** arises as a consequence of other types of underlying diseases. Also, it may be due to side effects of pharmacological treatments, such as corticosteroids. The processes that can generate this pathology include malnutrition, osteopathy of metabolic origin such as hyperparathyroidism, prolonged immobilization or cancer.

The first approach to osteoporosis treatment is from diet and increased physical activity. Vitamin D and calcium supplements are also suggested. Exercise stimulates osteoblasts, which respond positively to mechanical stimuli that activate their mechanisms to generate bone matrix from macanostransduction processes. When the weakening process of the microarchitecture is intensifying, it goes on to pharmacological treatment. Until not long ago, hormone replacement therapies using estrogens and progesterone were used. This treatment was based on the fact that estrogens decrease bone resorption, as mentioned above. But in observations subsequent to the implementation of this type of hormone replacement treatment, a large increase in cardiovascular disease and breast cancer was demonstrated.

In replacement of estrogen therapy, the use Selective Estrogen Receptor Modulators (SERMs) has been implemented. This group of drugs, including raloxifene and denosumab, can act as estrogen receptor agonists by binding to them in bone tissue. In this way, they behave as estrogen on the bone, but they do not cause adverse effects such as an increased risk of developing breast cancer.

Another type of pharmacological therapy for osteoporosis is related to the use of bisphosphonates such as alendronate and risedronate. Its effect is related to the inhibition of osteoclast activity, by inducing its apoptosis.

Recombinant parathyroid hormone is used as hormonal therapy for the treatment of osteoporosis. Its activity occurs at the level of the stimulation of osteoblast activity, and its effect is observed at the level of increased trabecular bone. It is also used siRNA and growth factors.

However, these treatments find many adverse effects. Prolonged use of raloxifene has been associated with the generation of venous thromboembolism while denosumab generates hypocalcaemia and cardiac abnormalities. Bisphosphonates cause gastritis and generate bone fractures after prolonged periods of treatment.

It is here where nanotechnology can provide many tools to specifically target drugs to bone, in order to increase selectivity and decrease adverse effects on other organs and tissues.

3.2 Bone Cancer

Tumorous bone diseases have two origins: primary and metastatic. Regarding primary bone tumors, osteosarcoma, chondrosarcoma and Ewing sarcoma represent 70% of malignancies. The common characteristic of this type of pathology is that they originate in mesenchymal stem cells and the stromal cells contribute to the growth of the tumor (Fig. 3.2).

3.2.1 Osteosarcoma

Sarcomas are the most common primary tumors. They are considered extremely invasive, generating a high mortality rate. They have a high incidence in children, adolescents and young adults: 70% of primary cases can be cured, but only 20% of osteosarcomas from metastases survive after 5 years from diagnosis.

Osteosarcoma finds origin in mesenchymal stem/stromal cells (MSCs) as well as in osteoblastic precursors.

The microenvironment generated in the tumor area is very complex, involving osteoblasts, osteocytes, osteoclasts, fibroblasts, immune, stromal and vascular cells and it is different and particular even between individuals.

The key point in osteosarcoma is related to the vicious circle between tumour cells and osteoclasts, in which tumor excessively stimulates osteoclast activity leading to degradation of bone matrix. This generates an osteolysis process that entails an early and painful fragility of the bone.

Research into strategies to address osteosarcoma is extensive. The problem is related to the to the complexity of the tumor, being very difficult to address all the parameters that govern its growth.

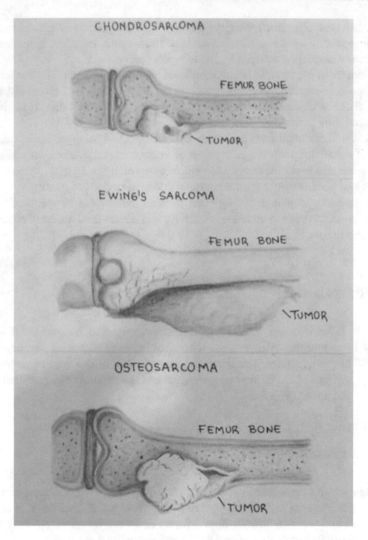

Fig. 3.2 Most common types of cancer that affect bone

In the late 1970s, chemotherapeutic treatment for patients with osteosarcoma was implemented. The procedure involves a chemotherapy treatment prior to surgery, and a subsequent one based on a cocktail of chemotherapeutic agents. One protocol is based on the use of high doses of methotrexate combined with etoposide and ifosfamide, which is used in children and adults under 25 years old. Other protocols combine also doxorubicin and cisplatin with above mentioned drugs.

Although these treatments are still in force, in the last 40 years the survival rate of affected patients has not been improved.

Many mechanisms are under investigation as different therapeutic strategies to counteract the growth and deleterious effects of osteosarcoma. In the last decade, a lot of research work has been devoted to determining the molecular bases of this pathology that allow for a more effective therapeutic approach.

This is important because the fact that there is no specific therapy is because many mechanisms related to the regulation of bone cell physiology are still being unveiled.

Being osteolysis the main problem associated with osteosarcoma, the first treatments were oriented to alleviate this aspect. Therefore, zoledronic acid, a bisphosphonate was investigated as anti-resorptive agent in combination with chemotherapy. However, clinical studied revealed no benefits.

The main physiological mechanism regulating the activity of osteoblasts and osteoclasts is the RANK/RANK-L system. Much effort has been directed to the investigation of the functioning of this system as a therapeutic target.

Demosumab is a humanized antibody against RANKL. This alternative was considered as promising to decrease the stimulation of osteoclastic activity. However, this strategy is not efficient considering that RANK signalling does not occur only at the osteoclast level. Thus, it is highly nonspecific and does not focus on point control of the division and excessive activity of osteosarcoma cells. In addition to this concept, it has been revealed that

> In last years, great research has been develop to improve the therapeutics related to osteosarcoma. The complex regulation of bone physiology and biology leads to think in several pathways to trigger therapeutically. In 2018 a study published by Ikebuchi et al., triggered a change in the paradigm regarding the conception of regulation by the RANK/RANK-L system. In this research work, osteoclasts were shown to secrete extracellular vesicles that expose RANK. They can interact with RANKL on the surface of osteoblasts. Through this interaction, an intracellular route would be activated in those related to Runx2 activation, stimulating the osteogenesis process. In this way, the RANK system involves the regulation of osteoclastic and osteoblastic activity simultaneously in order to balance said processes under physiological conditions.

All this information is novel and recent. Most of these therapeutic strategies are on clinical study. In this point, nanotechnology may emerge as a valuable tool to target chemicals and therapeutic agents to bone. In special, silica nanoparticles, considering the bioactive properties of silicon, as it will be described in coming chapters.

Two other types of tumor pathologies that affect the bone are chondrosarcoma and Ewin's sarcoma.

3.2.2 Chondrosarcoma

Chondrosarcoma is the second malignant tumor that originates from the bone, after osteosarcoma. It is characterized by malignant cells that produce a large amount of cartilage matrix. In general, they usually arises in long bones and pelvis in patients between 30 and 60 years old.

They are slow growing tumors. Due to their large amount of extracellular matrix and their low vascularity, they are very resistant to therapeutics as chemo or radiotherapy. Thus, surgery turns out to be the most effective treatment.

However, in those cases of tumor remission, these are usually much more malignant than the original tumor. Those cases where the tumor undergoes dedifferentiation tend to have a very poor prognosis. Different types of chondrosarcoma maybe described:

1. The chondrosarcoma is classified as conventional and its categorization is considered according to location: **primary chondrosarcoma** (75%) is generally located centrally in bone in the medullary cavity and **secondary chondrosarcoma** (10%), peripherally located at bone surface which are aggressive and present a very low survival estimated as 24% in 5 years. In this second case, most of tumors are the result of remission or malignant transformation. A minimum percentage of tumors estimated present peripheral location and they origin from periosteum, named as juxtacortical chondrosarcoma. It is important to distinguish histologically these two types of tumors, because juxtacortical chondrosarcoma present better prognosis after a correct surgery.

2. **Mesenchymal chondrosarcoma** occurs in the second and third decade of life and it is very aggressive with high risk of metastasis and local recurrence, being characterized by cartilaginous differentiation mixed with undifferentiated round cells. Tumors overexpress p53 without observation of mutations in most of cases studied. For this kind of tumor it has been observed expression of BCL2, platelet derived growth factor receptor (PDGFR) and protein kinase C (PKC).

3. **Dedifferentiated chondrosarcoma** are considered as hide grade tumors representing near 10% of chondrosarcoma. They are very aggressive and have a poor prognosis. This tumor may be observed as a noncartilaginous sarcoma which grow next to a malignant cartilage-forming tumor.

4. **Clear cell chondrosarcomas** (<2%) are tumors that appear in the epiphyseal end of long bones. They present good prognosis with correct resection. The histological characteristic of these tumor cells is that they present an empty cytoplasm. Something to take account with this tumor is that metastasis can occur after near 25 years, so long-term follow up is mandatory. It has been observed an over expression of PTHLH, PDGF, IHH, Runt related transcription factor 2 and matrix metalloproteinase 2. These characteristics may be considered for developing of specific therapies.

5. **Extraskeletal myxoid chondrosarcoma** (<2%) is considered as a differentiated tumor with myxoid degeneration. There exists contradictions in the conception of this kind of tumor and sometimes they are considered as two different entities: myxoid chondrosarcoma and extraskeletal myxoid "chondrosarcoma" (EMC). Therefore, the classification of this entity may lie as "tumors of uncertain classification". This was considered because the characteristic translocation (9:22) for extraskeletal myxoid chondrosarcoma is absent in bone myxoid chondrosarcoma.

Histology is a key tool in this type of tumor as it is a predictor for clinical behaviour. Also to take decisions in terms of surgery and therapeutics. Both central and peripheral chondrosarcoma are similar in histology. They can be classified into three grades.

It also exists a distinction between benign and malignancy of the tumors. Enchondroma and osteochondroma constitute benign tumors and are very difficult to differentiate from grade I chondrosarcoma, both histologically and radiologically. Due to this difficulty in differentiation, when faced with doubt, the same protocol is applied to both enchondroma and chondrosarcoma. Is consists in curettage, adjuvant use of phenol or cryosurgery.

In the following table, the different grades for chondrosarcoma are presented. In general, Grade I and Grade II are not differenced, considered all cases as Grade I.

Chondrosarcoma	Characteristics
Grade I	• Lowly cellular • Abundant hyaline cartilage matrix • Low possibility of metastasis
Grade II	• Difficult to differentiate from Grade I • Most quantity of cells in the matrix with atypical nuclear hyperchromasia • Similar prognosis than Grade I
Grade III	• Highly cellular • Mucomyxoid matrix with high grade of mitoses • High metastasis, up to 70%

Considering the difficulty in differentiating the stage, knowledge of the molecular biology of the tumor is extremely useful for the diagnosis to evaluate different markers that allow a more accurate diagnosis.

The molecular mechanisms that govern the differences between the origins of the different types of tumors related to cartilage are known, in special for chondrosarcoma.

In the last 4 years, there have been great advances in molecular studies related to chondrosarcoma, which are very useful for the application of specific treatments depending on the grade of the tumor. The most important is that they lay the foundations for potential specific treatments, and allow for accurate diagnosis and prognosis. This aspect is fundamental and many disciplines converge from research to clinic that allow a more specific approach to treatment.

Until 2017, diverse potential treatments were under study all including diverse strategies related to the different pathways involved in chondrosarcoma progression. These research is very important considering that there is no specific therapeutic strategies for this kind of neoplasm and also because of the difficulties associated to tumor targeting due to its resistance to drug entering. These strategies are resumed in the following table.

Target pathway	Characteristic	Proposed therapeutics
IDH1/IDH2	Proteins encoded by these genes catalyse oxidative decarboxylation of isocitrate in Kreb's cycle. This leads hypermethylation of DNA and histones causing tumorogenesis	Preclinical evaluation of IDH inhibitors
Hedgehog	Indian Hedhehog (IHH)/parathyroid hormone—related peptide pathway (PTH-P) is related to chondrocyte differentiation. Key role in pathogenesis of chondrosarcoma	Phase II trial of Saridegib, (IPI-926) resulted discouraging in advanced chondrosarcoma
Src	Regulates signalling transduction from surface receptors. Involves responses related to cell proliferation, migration, survival and also angiogenesis	Dasatinib is proposed as preclinical inhibitor of this pathway, alone or in combination with other drugs such as doxorubicin
PI3K-Akt-mTOR	Present in normal cells associated to growth, proliferation and survival. This pathway is activated by ligand interaction and involves IGF-1R, PDGFR-α, PDGFR-β	Preclinical and early phase trials indicate that inhibition of both mTORC1 and IGF-1R present better results
Angiogenesis	Associated to progression and metastasis of advanced chondrosarcoma, in special grade III. Includes VEGF, COX-2, VEGFR2, PDGFR-β, FGFR1	Angiogenesis inhibitors induce reduction of chondrosarcoma growth in cells studies and animal models. show tumor growth inhibition in chondrosarcoma animal models and cell lines
Histone deacetylation	Deacetylation of histones regulates gene expression. In general Through the deacetylation of histones, the DNA is transferred to a state of transcriptional silencing: chromatin is condensed and therefore the gene is not expressed	Romidepsin in phase II clinical trial as inhibitor of histone deacetylase induce apoptosis

The limitations associated to these treatments, which are currently under study, are because they target different mechanisms, which are common to diverse types of cancer, and they are not strictly specific to chondrosarcoma.

A current potential strategy, which is under research study for treatment of chondrosarcoma, is based on microRNAs. They are a group of noncoding RNAs which circulates in blood. They express in physiological conditions as well as in disease and they can be considered for diagnosis, prognosis and targeted treatment for tumoral bone diseases. It has been detected that in cases of chondrosarcoma, overexpression of miR-181a which is an oncomiR related to tumor progression. The mechanism involved in this pathway consists in the up regulation of protein-G signaling by miR-181a generating angiogenesis and metastasis. Recently, nanoparticles were developed for systemic treatment involving anti-miRNA oligonucleotides against miR-181a. It was observed that the implementation of this treatment in mice induced a decrease in the expression of mediators involved in the mentioned pathway such as VEGF and MMP1. The tumor volume was reduced in 32 after 38 days, prolonging survival from 23 to 45%. This is, at the time, the only research involving nanotechnology for this kind of pathology.

Within this line related to miRNAs, the active role of resistin in the regulation of signalling pathways that promote the progress of chondrosarcoma has recently been detected. Resistin is a kind of adipokine which has been detected to induce chondrosarcoma progression by stimulating the production of the angiogenic factor VEGF-A—The action of resistin is associated to the activation of phosphatidylinositol-3-kinase (PI3K) and Akt signaling pathways, which also involves the micro RNA miR-16-5p. This research revealed that downregulation of miR-16-5p induced angiogenesis by VEGF-A (Fig. 3.3).

With all the above, it is possible to observe that the treatment of chondrosarcoma continues to be a subject that still requires a lot of scientific and clinical research. Especially in regards to the inaccessibility of the tumor due to its characteristic matrix. In this aspect, nanotechnology finds a niche where its application could constitute a key and promising point for the development of effective therapies.

3.2.3 Ewing Sarcoma

Ewing sarcoma belongs to Ewing's sarcoma family tumors (ESFT) representing the second most common bone malignancy after osteosarcoma. Usually it occurs in children and young adults, with high at near 15 years old. The pelvis, the diaphysis of long bones and chest are the most common areas affected.

The origins of this type of tumor is genetically known. In the 85% of cases, it has been observed the t(11;22) (q24:q12) chromosomal translocation which induces a fusion of the 50 segment of EWS with the 30 segment of FLI-1 gene. The protein encoded by this translocation results a transcriptional activator that alters expression of genes. A multimodal approach is employed for the treatment, including chemotherapy, surgery and radiation. The prognosis for this pathology is very low,

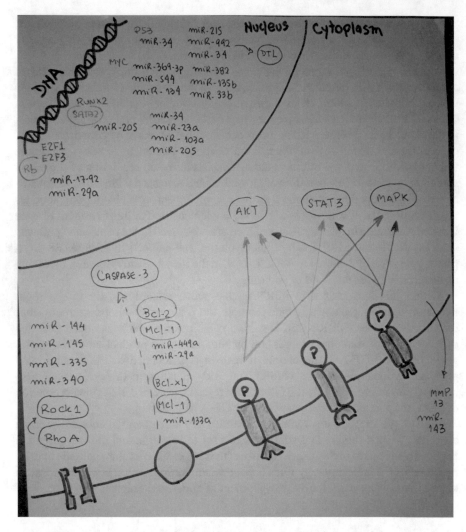

Fig. 3.3 Mechanisms based on miRNAs which govern molecular biology of chondrosarcoma

being 50% at 5 years, only 25% in the case of metastasis and less than 30% at 10 years.

3.3 Bone Infections

Bone infection frequently occurs after implantation of medical devices in bone. This is common in aged patients due to bone erosion and consequently fracture, but also occurs in a great part of population needing implants, either due to severe fractures or the need for post-surgical bone replacement associated with the removal of tumors. This issue constitutes a problem for the health system, both because of the complexity it generates in the quality of life of patients, and because of the economic issues entailing.

Implantable medical devices resent potential for bacterial contamination. The most common bacteria involved in bone infections are Staphylococcus epidermis, Staphylococcus aureus, Escherichia coli and Pseudomonas aeruginosa. The great problem associated with these infections is the formation of biofilm. They consist in a polysaccharide matrix excreted by microorganisms which growth in their own matrix inducing the formation of the biofilm. This king of infection is primarily due to the development of drug-resistant bacteria as a consequence of the inappropriate use of antibiotics. The treatment of biofilms is very difficult due to this resistance leading to uncontrollable growth protected by the inaccessible matrix. This matrix is resistant to the action the immune system as well as to the entry of therapeutic agents.

The low permeability of the matrix is not the only reason for biofilm resistance to antibiotics:

- Special characteristics of proteins produced by bacteria and also special characteristics of proteins belonging to the host.
- Bacteria with acquired resistance mechanisms: new enzymes with capacity of antibiotics degradation; development of efflux pumps; development of novel mechanisms of interactions with neighbouring bacteria which improve impermeability.

The biofilm formation, as illustrated below, involves a first step consisting in the adhesion of circulating bacteria to the surface of the medical device to be employed as implant. Then, bacteria start growing adopting a multilayer spatial organization. After maturation they start to produce the matrix originating the biofilm. The first step of this process is a key to avoid maturation and formation, so this window represents the opportunity for prevention. Nanotechnology may contribute to the development of nanomaterials devoted to avoid bacteria colonization (Fig. 3.4).

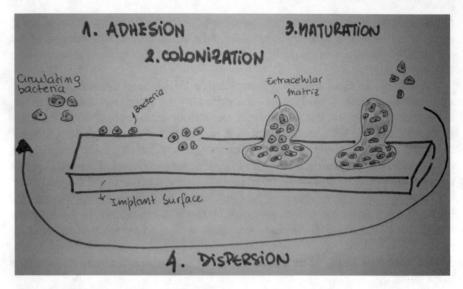

Fig. 3.4 Different states for biofilm formation

References

S. Alibert, J. N'gompaza Diarra, J. Hernandez, A. Stutzmann, M. Fouad, G. Boyer, J.M. Pagès, Multidrug efflux pumps and their role in antibiotic and antiseptic resistance: a pharmacodynamic perspective. Expert Opin. Drug Metab. Toxicol. **13**(3), 301–309 (2017)

D. Arcos, A.R. Boccaccini, M. Bohner, A. Díez-Pérez, M. Epple, E. Gómez-Barrena, A. Herrera, J.A. Planell, L. Rodríguez-Mañas, M. Vallet-Regí, The relevance of biomaterials to the prevention and treatment of osteoporosis. Acta Biomater. **10**(5), 1793–1805 (2014)

G. Bacci, S. Lari, Current treatment of high grade osteosarcoma of the extremity. J. Chemother. **13**(3), 235–243 (2001)

M.W. Bishop, K.A. Janeway, R. Gorlick, Future directions in the treatment of osteosarcoma. Curr. Opin. Pediatr. **28**(1), 26 (2016)

J.P. Brown, S. Morin, W. Leslie, A. Papaioannou, A.M. Cheung, K.S. Davison, D. Goltzman, D.A. Hanley, A. Hodsman, R. Josse, A. Jovaisas, Bisphosphonates for treatment of osteoporosis: expected benefits, potential harms, and drug holidays. Can. Fam. Physician **60**(4), 324–333 (2014)

S.S. Chen, C.H. Tang, M.J. Chie, C.H. Tsai, Y.C. Fong, Y.C. Lu, W.C. Chen, C.T. Lai, C.Y. Wei, H.C. Tai, W.Y. Chou, Resistin facilitates VEGF-A-dependent angiogenesis by inhibiting miR-16-5p in human chondrosarcoma cells. Cell Death Dis. **10**(1), 1–12 (2019)

M. Colilla, I. Izquierdo-Barba, M. Vallet-Regí, The role of zwitterionic materials in the fight against proteins and bacteria. Medicines **5**(4), 125 (2018)

C. Cooper, G. Campion, L. 3. Melton, Hip fractures in the elderly: a world-wide projection. Osteoporos. Int. **2**(6), 285–289 (1992)

P. D'Amelio, G.C. Isaia, The use of raloxifene in osteoporosis treatment. Expert Opin. Pharmacother. **14**(7), 949–956 (2013)

Y. de Jong, F. Bennani, J.G. van Oosterwijk, G. Alberti, Z. Baranski, P. Wijers-Koster, S. Venneker, I.H. Briaire-de Bruijn, B.E. van de Akker, H. Baelde, A.M. Cleton-Jansen, A screening-based approach identifies cell cycle regulators AURKA, CHK1 and PLK1 as targetable regulators of chondrosarcoma cell survival. J. Bone Oncol. **19**, 100268 (2019)

M.E. de Kraker, A.J. Stewardson, S. Harbarth, Will 10 million people die a year due to antimicrobial resistance by 2050? PLoS Med. **13**(11), e1002184 (2016)

M. Gaebler, A. Silvestri, J. Haybaeck, P. Reichardt, C.D. Lowery, L.F. Stancato, G. Zybarth, C.R. Regenbrecht, Three-dimensional patient-derived in vitro sarcoma models: promising tools for improving clinical tumor management. Front. Oncol. **7**, 203 (2017)

N. Gaspar, B.V. Occean, H. Pacquement, E. Bompas, C. Bouvier, H.J. Brisse, M.P. Castex, N. Cheurfa, N. Corradini, J. Delaye, N. Entz-Werlé, Results of methotrexate-etoposide-ifosfamide based regimen (M-EI) in osteosarcoma patients included in the French OS2006/sarcome-09 study. Eur. J. Cancer **88**, 57–66 (2018)

Y. Ikebuchi, S. Aoki, M. Honma, M. Hayashi, Y. Sugamori, M. Khan, Y. Kariya, G. Kato, Y. Tabata, J.M. Penninger, and N. Udagawa, Coupling of bone resorption and formation by RANKL reverse signalling. Nature **561**(7722), 195–200 (2018)

O. Johnell, J.A. Kanis, An estimate of the worldwide prevalence and disability associated with osteoporotic fractures. Osteoporos. Int. **17**(12), 1726–1733 (2006). 196. P. Mora-Raimundo, M. Manzano, M. Vallet-Regí, Nanoparticles for the treatment of osteoporosis

R.L. Jones, D. Katz, E.T. Loggers, D. Davidson, E.T. Rodler, S.M. Pollack, Clinical benefit of antiangiogenic therapy in advanced and metastatic chondrosarcoma. Med. Oncol. **34**(10), 167 (2017)

M. Kansara, M.W. Teng, M.J. Smyth, D.M. Thomas, Translational biology of osteosarcoma. Nat. Rev. Cancer **14**(11), 722–735 (2014)

A.D. Krol, A.H. Taminiau, J.V. Bovée, The clinical approach towards chondrosarcoma. Oncologist **13**, 320–329 (2008)

F. Limaiem, K.L. Sticco, Cancer, Chondrosarcoma, in *StatPearls* [Internet] (StatPearls Publishing, 2019)

P.A. Meyers, C.L. Schwartz, M. Krailo, E.S. Kleinerman, D. Betcher, M.L. Bernstein, E. Conrad, W. Ferguson, M. Gebhardt, A.M. Goorin, M.B. Harris, Osteosarcoma: a randomized, prospective trial of the addition of ifosfamide and/or muramyl tripeptide to cisplatin, doxorubicin, and high-dose methotrexate. J. Clin. Oncol. **23**(9), 2004–2011 (2005)

P. Mora Raimundo, M. Manzano García, M. Vallet Regí, Nanoparticles for the tratment of osteoporosis. AIMS Bioeng. **4**(2), 259–274.197 (2017)

A.J. Mutsaers, C.R. Walkley, Cells of origin in osteosarcoma: mesenchymal stem cells or osteoblast committed cells? Bone **62**, 56–63 (2014)

M. Nakagawa, F. Nakatani, H. Matsunaga, T. Seki, M. Endo, Y. Ogawara, Y. Machida, T. Katsumoto, K. Yamagata, A. Yamagata, S. Fujita, Selective inhibition of mutant IDH1 by DS-1001b ameliorates aberrant histone modifications and impairs tumor activity in chondrosarcoma. Oncogene **38**(42), 6835–6849 (2019)

B. Navet, K. Ando, J.W. Vargas-Franco, R. Brion, J. Amiaud, K. Mori, H. Yagita, C.G. Mueller, F. Verrecchia, C. Dumars, M.F. Heymann, The intrinsic and extrinsic implications of RANKL/RANK signaling in osteosarcoma: from tumor initiation to lung metastases. Cancers **10**(11), 398

S. Piperno-Neumann, M.C. Le Deley, F. Rédini, H. Pacquement, P. Marec-Bérard, P. Petit, H. Brisse, C. Lervat, J.C. Gentet, N. Entz-Werlé, A. Italiano, Zoledronate in combination with chemotherapy and surgery to treat osteosarcoma (OS2006): a randomised, multicentre, open-label, phase 3 trial. Lancet Oncol. **17**(8), 1070–1080 (2016)

G. Polychronidou, V. Karavasilis, S.M. Pollack, P.H. Huang, A. Lee, R.L. Jones, Novel therapeutic approaches in chondrosarcoma. Future Oncol. **13**(7), 637–648 (2017)

M. Spear, The biofilm challenge: breaking down the walls. Plast. Surg. Nurs. **31**(3), 117–120 (2011)

X. Sun, Y. Chen, H. Yu, J.T. Machan, A. Alladin, J. Ramirez, R. Taliano, J. Hart, Q. Chen, R.M. Terek, Anti-miRNA oligonucleotide therapy for chondrosarcoma. Mol. Cancer Ther. **18**(11), 2021–2029 (2019)

G. Taubes, The bacteria fight back. 356–361 (2008)

T. Tokatlian, T. Segura, siRNA applications in nanomedicine. Wiley Interdisc. Rev. Nanomed. Nanobiotechnol. **2**(3), 305–315 (2010)

C.G. Trejo, D. Lozano, M. Manzano, J.C. Doadrio, A.J. Salinas, S. Dapía, E. Gómez-Barrena, M. Vallet-Regí, N. García-Honduvilla, J. Buján, P. Esbrit, The osteoinductive properties of mesoporous silicate coated with osteostatin in a rabbit femur cavity defect model. Biomaterials **31**(33), 8564–8573 (2010)

R. Webb, K. Cutting, Biofilm the challenge. J. Wound Care **23**(11), 519 (2014)

S. Zaheer, M. LeBoff, E.M. Lewiecki, Denosumab for the treatment of osteoporosis. Expert Opin. Drug Metab. Toxicol. **11**(3), 461–470 (2015)

Chapter 4
Silicon: The Importance on Human Being and the Role on Bone Physiology

Many medical indications tend to motivate certain behaviours to preserve the health of the bone system. They have to do with eating a healthy diet and implementing physical activity. Regarding diet, it seems that calcium and vitamin D are the recommended protagonists to maintain or increase bone density, although there is not much evidence of these benefits. This may be due to two main issues: 1. more research may be needed; 2. calcium and vitamin D supplementation may not be enough and is also necessary to consider other nutrients such as magnesium, vitamin K, vitamin C, among others that play important roles in bone metabolism. For example, vitamin D and vitamin K act synergistically in the production and activation of osteocalcin.

At this moment, we welcome the leading role that silicon deserves in terms of its importance in bone metabolism and its health.

From a chemical point of view, silicon can be described as a metalloid found in-group 14 of the periodic table. After oxygen it is the most abundant element in the earth's crust and due to its great affinity for it, it is very rare to find it in its elemental form. For this reason it is commonly found naturally in oxidized form.

In nature there are various ways in which silicon can be found. Amorphous silica, SiO_2, is found in algae, sponges, and plants as part of the exoskeleton. In these cases it is also known as "biosilic" because it is biologically synthesized.

Orthosilicic acid is the soluble form in which silicon can also be found naturally: ($[Si(OH)_4]$) represents its monomeric form or in its hydrated phase as SiO_2 (H_2O) X, with suitable solubility and availability properties. This last hydrated form has more biological relevance since it can interact with organic molecules and also form complexes with compounds of inorganic nature.

Silicon is tetracoordinated in monomeric orthosilicic acid according to the following structure.

M. Agotegaray, *Silica-Based Nanotechnology for Bone Disease Treatment*, SpringerBriefs in Applied Sciences and Technology, https://doi.org/10.1007/978-3-030-64130-6_4

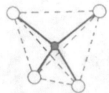

Although silicon plays a relevant role in the physiology of many tissues, such as skin, blood vessels and brain, we will focus our attention on bone.

Doing a little history it is worth mentioning the beginnings of the conception of silicon as one of the fundamental pillars in bone physiology. Edith Carlisle was the one who in 1970 worked with silicon in murine models and was able to highlight its role as the initiator of the mineralization process. Silicon was observed in high concentrations in the immature osteoid matrix and its level decreased as calcium began to deposit in the more mature matrix. Later, this statement was reinforced from a work where it was observed that the silicon supplement accelerated mineralization.

Deepening this research, Carlisle was able to relate, from a study carried out on chicks, that silicon plays an important role in the deposition of collagen in the matrix and in its subsequent calcification, regardless vitamin D levels.

By the 1970s, two other groups of researchers also began to work in relation to the role of silicon in bone physiology. Schwarz and Milne in 1972 followed by Nielsen and Sandstead in 1974 made important contributions on the essentiality of silicon for the development of the bone system.

Novel studies emerged in 1993 in which Hott et al. postulated the role of silicon as fundamental in an animal model of postmenopausal osteoporosis from finding that its supplement induced bone formation while minimizing its resorption.

It was from the year 2002, more attention has been paid to the relevant and active role that silicon plays in bone physiology.

From various clinical studies in patients with osteoporosis, it has been possible to establish a direct correlation between the levels of silicon, the formation of collagen and bone mineralization.

This trend can be established from various evidences that have been revealed from the study of cellular responses in relation to the proliferation or inhibition of osteoblasts and osteoclasts. Even, from deeper studies of the molecular mechanisms that are influenced with silicon supplements. Therefore, regarding evidence in experiments where the response in relation to cell proliferation has been evaluated, it has been observed that silicon stimulates the proliferation of MG63 osteoblasts.

On the other hand, the inhibitory role of silicon in osteoclastic activity has been evidenced in several studies. In this sense it can be from direct mechanisms or indirect from the interaction that exists between osteoblasts and osteoclasts. This observation was reinforced by a study in which the presence of silicon generated an increase in osteoprotegerin in cells similar to osteoblasts, which counteracted the catabolic role of the RANKL receptor, involved in the activation of osteoclasts.

Orthosilicic acid is closely related to different molecular mechanisms that govern the stimulation of osteogenesis.

- **Regulation of osteogenic gene expression**: Silicon has been shown to promote increased expression of various genes such as that encoding bone morphogenetic protein- (BMP2-2) and various transcription factors such as Runx-2 (Run related transcription factor 2) which is a transcription factor involved in the control of gene expression related to skeletal physiology and also the well-known transforming growth factor-β (TGF-β). Orthosilicic acid regulates gene expression of alkaline phosphatase and osteocalcin according to studies conducted on mRNA of osteoblasts derived from human bone.

- **Production of type I collagen**: It has been shown from different research that silicon is closely related to the stimulation of type I collagen production. The first studies carried out in this regard showed that silicic acid stimulates the prolyl hydroxylase activity that is involved in the synthesis of collagen. In this way, an increase in hydroxyproline levels has even been observed in tibiae of mice supplemented with silicon. In more detail, a decrease in the activity of the enzyme ornithine transaminase that participates in the synthesis of proline in the liver has been observed. In MG63 osteoblasts, the expression of the gene that codes for type I collagen was increased by regulating ERK kinase activity from the extracellular medium. Furthermore, silicon plays a very important role in the collagen assembly process and its mineralization. Thus, low concentrations of silicon induce alterations in the self-assembly process of the collagen subunits.

- **Biomineralization**: In addition to the proven relationship between the presence of silicon and the stimulation of the expression of genes related to osteogenesis and the activation of pathways involved in the activation of osteoblasts, this mineral plays a crucial role in the bone biomineralization process. In early calcification processes, its $Si(OH)_4$ form can induce hydroxyapatite precipitation as it has been observed in studies in electrolyte solutions.

- **RANK and OPG**: It has been observed in vitro that $Si(OH)4$ induces an increment in the expression of OPG without affecting RANK-L. In addition, it may bind to OPG, sequestering RANK-L. By this indirect mechanism pre-osteoclast maturation and the activation of osteoclasts would be negatively regulated.

The following figure summarizes the mechanisms by which silicon plays a crucial role in the process of osteogenesis (Fig. 4.1).

As mentioned above, the species in which silicon is mostly found in humans is orthosilicic acid $Si(OH)_4$. When ingested, it is absorbed at the intestinal level through transporters. Then, it passes to the blood compartment where it travels without joining other species although it can sometimes interact with iron and aluminium species.

The serum concentration can normally be between 24 and 31 g/dL. Once it passes into the tissues, it binds to glycosaminoglycae. The regulation between serum levels and those found in tissues is governed by transporters that have been found in plants and in organisms whose skeleton is mainly composed of silicon, such as diatoms or sponges. In mammals a silicon transporter called SLc34a2 has been revealed.

The structure of these transporter systems would be similar to aquoporin-type channels, the presence of which has been detected in kidney tissue, the small intestine,

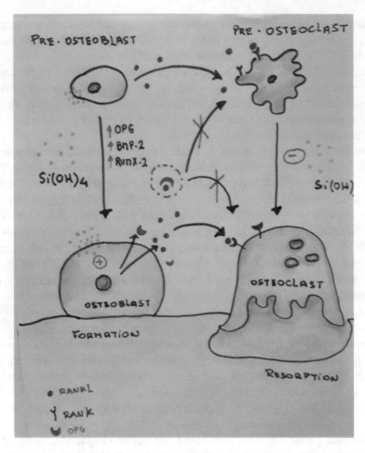

Fig. 4.1 Regulation of mechanisms involved in bone formation and resorption by Si(OH)$_4$

and bone, and their expression is governed according to the amounts of silicon. Silicon excretion occurs through urine through renal function and the evaluation of this parameter is an indicator of bone metabolism. Thus, a decreased urine silicon concentration could be an indicator of osteopenia.

In Fig. 4.2 there are foods rich in silicon whose intake provides the necessary amounts of this mineral, established between 10 and 25 mg daily.

Fig. 4.2 Main sources of silicon

References

M.Q. Arumugam, D.C. Ireland, R.A. Brooks, N. Rushton, W. Bonfield, The effect orthosilicic acid on collagen type I, alkaline phosphatase and osteocalcin mRNA expression in human bone-derived osteoblasts in vitro, in *Key Engineering Materials*, vol. 309 (Trans Tech Publications Ltd, 2006), pp. 121–124

M.C. Brady, P.R.M. Dobson, M. Thavarajah, J.A. Kanis, Zeolite A stimulates proliferation and protein synthesis in human osteoblast–like cells and osteosarcoma cell line MG-63. J. Bone Miner. Res. **6**, 39 (1991)

E.M. Carlisle, Silicon: a possible factor in bone calcification. Science **167**(3916), 279–280 (1970)

E.M. Carlisle, A relationship between silicon and calcium in bone formation, in *Federation Proceedings*, vol. 29, No. 2, p. A565 (1970, January). 9650 Rockville PIKE, Bethesda, MD 20814-3998: Federation of American Societies for Experimental Biology

E.M. Carlisle, Silicon: a requirement in bone formation independent of vitamin D 1. Calcif. Tissue Int. **33**(1), 27–34 (1981)

T. Gao, H.T. Aro, H. Ylänen, E. Vuorio, Silica-based bioactive glasses modulate expression of bone morphogenetic protein-2 mRNA in Saos-2 osteoblasts in vitro. Biomaterials **22**(12), 1475–1483 (2001)

J.R. Henstock, L.T. Canham, S.I. Anderson, Silicon: the evolution of its use in biomaterials. Acta Biomater. **11**, 17–26 (2015)

M. Hott, C. de Pollak, D. Modrowski, P.J. Marie, Short-term effects of organic silicon on trabecular bone in mature ovariectomized rats. Calcif. Tissue Int. **53**(3), 174–179 (1993)

H.M. Macdonald, A.C. Hardcastle, R. Jugdaohsingh, W.D. Fraser, D.M. Reid, J.J. Powell, Dietary silicon interacts with oestrogen to influence bone health: evidence from the Aberdeen Prospective Osteoporosis Screening Study. Bone **50**(3), 681–687 (2012)

F. Maehira, Y. Iinuma, Y. Eguchi, I. Miyagi, S. Teruya, Effects of soluble silicon compound and deep-sea water on biochemical and mechanical properties of bone and the related gene expression in mice. J. Bone Miner. Metab. **26**(5), 446 (2008)

W.E.G. Müller, X.H. Wang, F.Z. Cui, K.P. Jochum, W. Tremel, J. Bill, H.C. Schröder, F. Natalio, U. Schloßmacher, M. Wiens, Sponge spicules as blueprints for the biofabrication of inorganic–organic composites and biomaterials. Appl. Microbiol. Biotechnol. **83**, 397–413 (2009)

F.H. Nielsen, H.H. Sandstead, Are nickel, vanadium, silicon, fluorine, and tin essential for man? A review. Am. J. Clin. Nutr. **27**(5), 515–520 (1974)

A.M. Pietak, J.W. Reid, M.J. Stott, M. Sayer, Silicon substitution in the calcium phosphate bioceramics. Biomaterials **28**(28), 4023–4032 (2007)

A.A. Poundarik, T. Diab, G.E. Sroga, A. Ural, A.L. Boskey, C.M. Gundberg, D. Vashishth, Dilatational band formation in bone. Proc. Natl. Acad. Sci. **109**(47), 19178–19183 (2012)

C.T. Price, J.R. Langford, F.A. Liporace, Essential nutrients for bone health and a review of their availability in the average North American diet. Open Orthop. J. **6**, 143 (2012)

S. Ratcliffe, R. Ratcliffe, J. Vivancos, A. Marron, R. Deshmukh, J.F. Ma, N. Mitani-Ueno, J. Robertson, J. Wills, M.V. Boekschoten, M. Müller, Identification of a mammalian silicon transporter. Am. J. Physiology-Cell Physiol. **312**(5), C550–C561 (2017)

D.M. Reffitt, N. Ogston, R. Jugdaohsingh, H.F.J. Cheung, B.A.J. Evans, R.P.H. Thompson, J.J. Powell, G.N. Hampson, Orthosilicic acid stimulates collagen type 1 synthesis and osteoblastic differentiation in human osteoblast-like cells in vitro. Bone **32**(2), 127–135 (2003)

H.C. Schröder, X.H. Wang, M. Wiens, B. Diehl-Seifert, K. Kropf, U. Schloßmacher, W.E.G. Müller, Silicate modulates the cross-talk between osteoblasts (SaOS-2) and osteoclasts (RAW 264.7 cells): inhibition of osteoclast growth and differentiation. J. Cell. Biochem. **113**(10), 3197–3206 (2012)

K. Schwarz, D.B. Milne, Growth-promoting effects of silicon in rats. Nature **239**(5371), 333–334 (1972)

C.D. Seaborn, F.H. Nielsen, Silicon deprivation decreases collagen formation in wounds and bone, and ornithine transaminase enzyme activity in liver. Biol. Trace Elem. Res. **89**(3), 251–261 (2002)

M.Y. Shie, S.J. Ding, H.C. Chang, The role of silicon in osteoblast-like cell proliferation and apoptosis. Acta Biomater. **7**(6), 2604–2614 (2011)

S. Sripanyakorn, R. Jugdaohsingh, R.P. Thompson, J.J. Powell, Dietary silicon and bone health. Nutr. Bull. **30**(3), 222–230 (2005)

M. Wiens, X. Wang, U. Schloßmacher, I. Lieberwirth, G. Glasser, H. Ushijima, H.C. Schröder, W.E. Müller, Osteogenic potential of biosilica on human osteoblast-like (SaOS-2) cells. Calcif. Tissue Int. **87**(6), 513–524 (2010)

M. Wiens, X. Wang, H.C. Schröder, U. Kolb, U. Schloßmacher, H. Ushijima, W.E. Müller, The role of biosilica in the osteoprotegerin/RANKL ratio in human osteoblast-like cells. Biomaterials **31**(30), 7716–7725 (2010)

Chapter 5
Amorphous Silica Nanoparticles: Biocompatibility and Biodistribution

Once the anatomical, physiological and biochemical characteristics of the bone have been described and also, after having discussed the important role of silicon in bone physiology, now is the time to delve into the nanotechnology associated with silica nanoparticles as therapeutic agents for bone.

The use of nanotechnologies in the biomedical field and in the pharmaceutical industry has become extremely interesting and promising in recent years, especially with regard to diagnosis and the directed and controlled transport of drugs.

Nano/microparticle-based drug transport systems have many advantages from a therapeutic point of view:

(i)   The ability to be directed to specific sites in the body.
(ii) The reduction of the amount of drug necessary to reach a certain concentration in the target organ.
(iii) The decrease in concentration in unwanted sites, among others.

These characteristics contribute to the reduction of the adverse effects associated with certain drugs. All these properties justify the increase in the number of publications associated with the application of nanoparticles (NPs) as drug transport agents. There are various formulations, such as dendrimers, micelles, emulsions and liposomes as agents to reach a specific blank site in the body.

The NPs must have certain characteristics in order to be used as drug-therapeutic systems, which means that an adequate combination must be achieved between the size, the method to incorporate the drug of interest (adsorption or encapsulation), the surface chemical properties, hydrophilicity/hydrophobicity, surface functionalization, biodegradability, and physical properties (such as response to temperature, pH, electrical charge, magnetism).

In the last decade, the trend in biomedical and pharmacological applications of nanotechnology points to the development of comprehensive systems that not only allow the specific transport or targeting of drugs, but also facilitate the diagnosis and monitoring of nano-devices in vitro, initially for research and development

© The Author(s), under exclusive license to Springer Nature Switzerland AG 2020   45
M. Agotegaray, *Silica-Based Nanotechnology for Bone Disease Treatment*,
SpringerBriefs in Applied Sciences and Technology,
https://doi.org/10.1007/978-3-030-64130-6_5

issues, and then in vivo for the clinical realization of the desired applications. In this way, the "theranostic" discipline has emerged as an interdisciplinary field integrating image and therapy. Thus, the development of new nanotechnology platforms for both purposes results in one of the great challenges for cutting-edge biomedicine, where previously, during treatment, the monitoring of NPs could be used to monitor therapy.

Currently, the clinical detection of various pathologies, including bone-associated diseases, is based on diagnostic imaging or morphological analysis of suspicious cells (cytology) or tissues (histopathology) post-surgery. Imaging includes X-ray radiography, computed axial tomography, magnetic resonance imaging, and ultrasound. Although these are methods in force in medicine, they find some limitations in the ability to differentiate benign from malignant lesions in tumor pathologies.

Likewise, the limitations of cytological techniques are that they often do not have the necessary sensitivity to detect stages of early disease. Meanwhile, histopathological ones always require the accompaniment of the image by tomography or magnetic resonance imaging to carry out a good correlation and the problem is that it is not always possible to identify early stages of the disease in the images. Thus, the development of arrangements and methods for the early detection of various pathologies, before they become symptomatic, results in a great challenge. In this context, nanotechnology has tools with unique properties for the early diagnosis of various diseases, especially cancer. It is here where the development of NPs with fluorescent properties would constitute a valuable tool for early diagnosis for diseases. In turn, their concomitant use as drug transporters would lead to being able to implement early treatments that improve survival and quality of life of affected patients, while being able to monitor the evolution of the disease.

Fluorescence is an optical phenomenon where the absorption of photons at a certain wavelength results in the emission at another wavelength, usually greater. The energy loss between the absorbed and emitted photons is the result of the vibrational relaxation phenomenon, the difference of which refers to the Stokes displacement. In the first phase, known as excitation, light absorption results in the promotion of an electron from the ground state to an excited state. Once excited, the release of absorbed energy can occur through various photophysical events, including both radioactive and non-radioactive emission. Vibrational relaxation is often the first pathway for energy dissipation and can be followed by internal conversion, cross-systems (from a singlet to a triplet state) and subsequent phosphorescence or fluorescence when the excited electron returns to the ground state and emits energy through the release of a photon. Fluorescence spectroscopy is a very useful technique for detecting biomolecules and is widely used in biological and biomedical applications due to high spatial and temporal resolution. Any material with fluorescent detection properties can be used in a variety of analysis methods, including solution assays that employ fluorescence spectroscopy, fluorescence microscopy for cell imaging, flow cytometry for high-quality imaging of individual cells and even live images.

The use of fluorescence as a detection method depends on the photophysical properties of the fluorophore: photostability, quantum yield, Stökes shift and fluorescence time.

Silicon dioxide, SiO_2, a material known as "silica" has certain properties that make it a good candidate as a biomaterial: it has a very labile surface to be functionalized, ease of synthesis and biocompatibility.

Silica is presumed to be a non-toxic material in vivo, in concentrations as high as 50,000 ppm according to results obtained in rats.

There is much information in the scientific literature related to the development of mesoporous silica NPs for the controlled transport and release of drugs. They consist of structures where SiO_4 tetrahedral arrays form porous structures that allow drug loading.

The latest trends related to this type of materials are based on the combination of different strategies to obtain multifunctional drug transport systems where each device has specific properties regarding targeting and response to different stimuli such as temperature, pH, etc. so that the loaded drug is not released except at the specific site. Although there is a lot of information regarding the synthesis and characterization of these mesoporous silica-based nanosystems, much research is still needed on their toxicological impact in vivo regarding the correlation between size, pore structure, biodegradability and biocompatibility in living organisms to achieve a concrete implementation at the clinical level.

In relation to this type of bone-oriented nanoparticles, it has recently been published by Gisbert-Garzarán et al. (2020) a very comprehensive review describing the formulations developed so far for the treatment of osteoporosis, bone cancer and bone infections.

However, regarding amorphous silica as material with biomedical applications, there is not much development in research so far. The pioneering studies of Stöber and collaborators established the foundations for the world of nanoparticulate silica by using sol-gel chemistry to obtain spherical silica nanoparticles. The synthesis involves the dissolution of silicon alkoxides (tetraethyl orthosilicate, TEOS) in hydroalcoholic solution in basic medium, provided by ammonia. From this study, much research work has been carried out to introduce modifications to the original method with the aim of improving the quality of the silica NPs obtained and to functionalize the surface of the same according to the desired applications. All the variables introduced by Stöber were meticulously studied by Rao et al. (2005), revealing that the size of the particles depends on the type of alkoxide used, the concentration of alcohol and alkali, as well as the conditions of synthesis such as temperature and ultrasound application. Subsequently, Rahman et al. (2007) continued to work on optimizing the synthesis conditions from the introduction of modifications to the Stöber method. In this case, they used the same precursors, varying the concentration as a study variable, under the influence of ultrasound, just like Rao et al., but introduced a modification based on the calcination of the NPs obtained at 500 °C with re-hydroxylation from exposure of the sample to air for 60 days. The authors observed an increase in the number of silanol groups in agreement with the decrease in size, especially below 40 nm. In any case, the synthesis method is long and tedious if you think about scaling to reach specific applications. The design of the surface of silica NPs implies an adequate balance in the use of inert and active functional groups with the aim of minimizing the aggregation of the NPs, reducing nonspecific binding and in turn imparting anchor sites for the drugs of interest if they are used for pharmacological applications.

In this context, Bagwe et al. (2006) prepared silica NPs by the microemulsion method and then modified them with TEOS and different organosilanes by the co-hydrolysis method to obtain nanosystems functionalized with carboxylates, amines, amine/bisphosphonate, polyethylene glycol (PEG), octadecyl and carboxylate/octadecyl groups. To determine the effect of different surface modification patterns on nonspecific junctions, the authors doped the surface with a fluorescent dye and determined that functionalization facilitates interaction with biomolecules. For their part, previously, Xu et al. (2002) obtained spherical and monodisperse PEG-coated silica NPs, with sizes between 50 and 350 nm, using the tetramethylortosilicate hydrolysis (TMOS) method. The PEG coating helps to reduce the adsorption of proteins on the surface of the NPs, thus increasing their biocompatibility.

Considering the potential properties in pharmacological terms of these nano-devices and with the promising emergence of the teranostic field, much work was carried out based on the incorporation of fluorophores into the silica NPs conventionally obtained by the Stöber method by doping or entrapment (Van Blaaderen and Vrij 1992), since they are not naturally luminescent. Thus, organic or inorganic fluorophores have been incorporated into the nanoparticulate silica structure through non-covalent interactions (Chinen et al. 2015). It has been shown that the fluorophores associated with the silica NPs are less toxic than the free fluorophores and that, in addition, the external surface of the NPs is plausible to modify with functional groups that allow the subsequent anchoring of various biomolecules (Wang et al. 2011; Xue et al. 2010). The disadvantage of this type of nano-systems is that they are not useful systems for long-term treatments since they have a tendency to release trapped dyes, thus reducing the luminescent capacity and limiting their application.

Fluorescent NPs by themselves would offer enhanced properties in relation to these properties with respect to conventional organic fluorophores. In this context, post-obtaining calcination of silica NPs has been shown to yield luminescent nano-systems. The origin of the luminescence phenomenon in these modified silica NPs is due to the presence of defects in the silica centers, to carbon and oxygen impurities and to charge transfer mechanisms, among others (Yoldas 1990; Garcia et al. 1995; Green et al. 1997) in the silica network resulting from the calcination of the amino-propyl groups. Obtaining various luminescent silica nano-systems has been reported. Jakob et al. (2006) reported the synthesis of monodisperse luminescent silica spheres obtained by calcination of aminopropyl-silica NPs synthesized from the mixture of TEOS and aminopropyltriethoxysilane (APTES) in a solution composed of ammonium hydroxide, water and ethanol (components of the method from Stöber). The optical properties were different, depending on the calcination temperature and the TEOS: ammonium molar ratio. Then, Brites et al. (2012) reported obtaining luminescent silica NPs at low temperature (45 °C) from the Stöber method using TEOS and APTES as precursors. The latest work to obtain luminescent silica NPs is very recent (Chandra et al. 2017) and reports a one-step synthesis method of fluorescent NPs functionalized with amino-terminal groups, using silicon tetrabromide as precursors

(SiBr$_4$) and APTES. The NPs have sizes between 1 and 2 nm in diameter and exhibit intense blue light emission with a quantum yield of 34% in aqueous medium.

Silica NPs would be a promising device as carriers of drugs to the bone since they would have the property not only of transporting drugs intended for bone therapy, but would also contribute to the bone remodelling process.

Considering the above, NPs containing silica in their composition would be a promising device as carriers of drugs to the bone since they would have the property not only of transporting drugs intended for bone therapy, but would also contribute to the bone remodelling process. Silica in the form of NPs has been postulated as a bioactive and beneficial material for bone.

When NPs are oriented to biomedical applications, it is crucial to know their biocompatibility profile in order to ensure their usefulness. There is much controversy today regarding the toxicity of amorphous silica nanoparticles. Many authors maintain the concept of their safety while others argue that they would be toxic. In general, their administration in vivo is carried out parenterally so that, as with any foreign material, the organism will respond to its entry. In a very recent and detailed review of the toxicological aspects related to silica NPs, Murugadoss et al. (2017) conclude that the study of surface functionalizing agents and the correlation of physicochemical properties with toxicity is crucial to be able to draw conclusions regarding toxic effects. Thus, further research regarding in vitro and in vivo interactions, bioavailability and bioaccumulation is required to understand the toxicity of these nanosystems.

Diverse factors determine the toxicity of amorphous silica nanoparticles:

- Size
- Surface charge
- Surface functionalization
- Porosity
- Type of cell involved in the interactions.

If NPs are intended for biomedical applications and little is known about their toxicological effects both in vitro and in vivo, it is useful to consider the pathways of entry into the body and the organs and tissues, which NPs will be found acutely. Upon entering the bloodstream, the NPs will surround themselves with plasma proteins and will interact with the formed elements of the blood before passing through the endothelium to reach the organs and tissues completing the biodistribution process. Once in contact with biological matrices, the first process that NPs undergo is the coating with proteins, giving rise to what is called "protein corona" (Lynch and Dawson 2008). This coating is complex and variable considering that the complete plasma proteome contains around 3700 proteins. Fifty of them 50 have been identified in association with NPs. Protein adsorption depends on various factors, among which are the size and surface characteristics of NPs and this process is a determining factor in biodistribution in addition to the inherent physicochemical properties of NPs, such as size, shape, surface charge, solubility. The mechanism of binding of plasma proteins to NPs has not been fully elucidated so far. To understand the mechanism of formation of the corona protein around a specific nanosystem, it is not only necessary to know which proteins interact, but also the kinetics of the interaction, the affinity and the stoichiometry of association-dissociation protein versus nanoparticles.

The study of protein corona formation on amorphous silica nanoparticles is relatively recent.

On this subject, it is known that the surface of the silica nanoparticles is decisive in the formation of the corona protein. In addition, size and porosity are also two determining properties. The formation of protein corona conditions cellular uptake and biocompatibility.

In this regard, it is important to mention that the corona protein generates a coating that would increase biocompatibility and also induces cell uptake without membranolysis. This was observed in lung cells exposed to nanosilica particles considering that this kind of nanomaterials was observed to induce pulmonary toxicity. In this case, pro-inflammatory factors were observed to be reduced in the case of cells treated with protein corona-modified silica surface in contrast to bare nanosilica. In addition, this protein coating facilitated uptake.

In contrast, in previous research performed by it was observed that addition of 10% serum to culture medium of epithelial lung cells incubated with 50 nm amorphous silica nanoparticles did not facilitate nanomaterial uptake.

It has been observed in vitro that naked silica nanoparticles can cause alterations in cell membranes by direct contact. This effect is size and dose dependant. In a recent study performed on erythrocytes, lymphocytes, macrophages and malignant melanocytes, it has been demonstrated that the addition of serum proteins to the culture medium would act, after interacting with the surface of the nanoparticles, as a protector. This would even favour interaction with the cell membrane without membranolysis.

As new works appear, new information is obtained about the behaviour of amorphous silica nanoparticles in terms of their cytotoxicity and their interaction with cells. For example, a study performed on amorphous silica nanoparticles differently functionalized with amine and carboxylate groups, and also modified or non-modified with polyethylene glycol (PEG) gave new insights in terms of data regarding formation of protein corona as well as cell interactions. It was observed that initial positive surface charge facilitates interaction with macrophages. Thus, Kurtz-Chalot et al. (2017) found that in vitro cytotoxicity was not in direct relation with uptake of nanoparticles and PEGylating prevents uptake. They found a correlation with composition of protein corona and initial surface functionalization of nanoparticles on macrophages. In this case, the identity of protein corona influenced by initial surface charge was a key factor on cell interactions and thus, uptake and cytotoxicity.

Unlike many works where they emphasize that amorphous silica nanoparticles would exert their cytotoxic effects by generating oxidative stress, in this work this mechanism was not evidenced in macrophages.

These are some research work, not all, leading with different results from studying cytotoxicity on different cell types of different silica nanoparticles in terms of surface charge, size and different conditions related to interactions with proteins.

The discrepancies observed among different works may be explained by several hypothesis being three of them the most relevant:

- Surface of nanoparticles determines composition of protein corona
- Presence or absence of protein corona influences on nanoparticles-cell interactions
- Uptake and cytotoxicity would depend on the internalization pathway and on intrinsic mechanisms expressed on each cell type.

The protein corona as well as the size of the nanomaterial may trigger different uptake mechanisms, either specific involving receptors on plasmatic membrane or non-specific pathways. For example, phagocytosis for larger nanoparticles, micropinocytosis for fluid uptake, and endocytic pathways mediated by clathrin or caveolin-mediated endocytosis.

It has recently been shown that it is possible to modify the surface of nanoparticles with agents that prevent from protein corona formation. This is important to reduce the gap between the functionality that is intended to be imparted from the design and engineering and the biological effects observed in vitro and what actually occurs when nanoparticles interact with the complex media they encounter in the body.

The binding of proteins to silica nanoparticles surface induces an increase in size and changes in zeta potential values, which is correlated with surface charge. Several studies emphasize the direct relation between toxicity and size: larger size of amorphous silica nanoparticles would contribute to an increase in biocompatibility, but this is a general trend taking into account that several authors indicated the contrary.

It is not easy to study the composition of protein corona due to the great composition of plasma in terms of proteins. The conventional technique employed for determination of hydrodynamic diameter and size is "Dynamic light scattering" (DLS). This is very useful for the study of nanoparticles in simple medium. But in the cases of serum or physiological simulating media, aggregations or agglomerations of nanoparticles, DLS is no proper. Tunable Resistive Pulse Sensing arises as useful technique as well as Electrophoretic Light Scattering in order to determine zeta potential measurements of protein corona-modified silica nanoparticles. But these techniques only give information in relation to surface charge; not in terms of composition.

Proteomic analysis of protein corona composition is possible employing specific techniques such as SDS-PAGE gel electrophoresis (sodium dodecyl sulphate polyacrylamide gel electrophoresis). After separation of the different bands of interest and posterior digestion, composition can be analysed by electrospray liquid chromatography mass spectrometry (LC MS/MS) using an HPLC. Proper databases are needed depending on the proteins employed to study protein corona.

All this information recovered shows that it is difficult to correlate the toxicity of amorphous silica nanoparticles either with their size, their surface charge or their composition. Rather we can make clear from this analysis that it would be important to carry out the toxicological study of each specific nanosystem with respect to the cell type or tissue with which it is desired to establish an effect. And in this way we can talk about precision biomedicine for silica nanoparticles. Since obviously many factors govern cytotoxicity, and involve intrinsic questions of nanoparticles and questions related to the biology of the cell type in question.

The application of NPs as biomedical agents, in special for drug targeting, depends largely on that they are not rapidly eliminated from the bloodstream by cells of the immune system. As previously mentioned, when NPs enter the blood circulation they find a complex environment of plasma proteins and cells of the immune system. NPs tend to be captured both in circulation by monocytes, platelets, leukocytes, and dendritic cells and in tissues by resident phagocytic cells (Kupffer cells in the liver, dendritic cells in lymph nodes, and B cells in the spleen) (Cedervall et al. 2007). The clearance of NPs depends on the characteristics of the corona protein and determines its circulation time and the number of particles available to reach the blank site.

The fraction of erythrocytes in the blood is much greater than the fraction occupied by mononuclear phagocytic cells, thus, it is more likely that when entering the blood, the nanoparticles interact more with them than with cells of the immune system. The study of the induction of hemolysis or the generation of morphological changes of the red blood cells from the contact with the nanodevices is a very important aspect within their pre-clinical characterization. Currently, there are very few studies devoted to evaluating the impact of silica NPs on blood cells, in special, erythrocytes.

In general, silica nanoparticles induce hemolysis associated to different mechanisms. In addition, it has been observed that this effect is size and dose-dependent. Yu et al. (2011) revealed that hemolysis is highly dependent on surface characteristics. The functionalization of the surface with amino groups would decrease the hemolytic effect of the NPs, although this effect would be dependent on the concentration of surface groups, since an excess of them would generate the same effect as the non-functionalized NPs.

Smaller silica nanoparticles result as more toxic for erythrocytes. The interaction of 50 nm sized amorphous silica nanoparticles was studied at different concentrations from 1 to 125 μg/ml by Nemmar et al. In this work, it was found an increase toxicity related to increasing doses. Different parameters were found to be exacerbated such as lactate dehydrogenase activity, in vitro lipid peroxidation, superoxide dismutase and catalase activity accompanied by a reduction in glutathione. These parameters indicate oxidative stress conditions. In addition, different parameters such as an increase in cytosolic calcium concentration associated to increment in caspase 3 was indicative of apoptosis induction. In this work it was also demonstrated the uptake of silica nanoparticles by erythrocytes.

Another study also concluded that amorphous silica NPs of around 60 nm cause dose-dependent hemolytic effects in human erythrocytes through mechanisms associated with oxidative stress and disorders of cell energy metabolism (Jiang et al. 2016).

Nemmar et al. (2015) also performed a study on platelets and found the same trend than for erythrocytes: increasing concentrations of silica nanoparticles rendered platelet aggregation, oxidative stress and increased level of intracellular calcium.

In vitro studies carried out on peripheral lymphocytes showed that the coating of the nanoparticles favors their biocompatibility, that the nanoparticles can be internalized. On the other hand, the bare surface of silica nanoparticles can lead to alterations of the cell cile in these cells. These effects were dose and size dependent.

A similar trend was observed on peripheral mononuclear cells, where cytotoxicity was size dependent observing more deleterious effects for smaller nanoparticles. This cytotoxicity was also associated to mechanisms related to oxidative stress.

All these cytotoxic trends may be avoided working deeply on the surface of silica nanoparticles. This is a perspective for future studies considering the versatility of their surface in terms of functionalization to obtain biocompatibility.

Once in contact with the blood, the NPs meet the endothelial wall to be transported to each of the body's organs and tissues. In this context, the study of the effects on endothelial cells is important to define aspects of their biocompatibility.

In a study performed by Napierska et al. (2009) on human endothelial cells, 15 nm silica nanoparticles exerted more toxic effect leading to cell death than nanoparticles of near 100 and 300 nm. In this work shape, surface charge and density of the nanoparticles dispersion also seemed to play an important role in toxicity.

It has been demonstrated that amorphous silica nanoparticles can penetrate endothelial cells. The effects generated inside the cell in terms of cytotoxic effects were observed to be related to size and concentration. Thus, smaller nanoparticles at high concentrations trended to induce an increase production of NO, generating a cascade of $ONOO^-$. The imbalance between NO and $ONOO^-$ may trigger inflammatory and cytotoxic responses. These effects were observable in the case of nanoparticles lower than 50 nm.

In a study developed by Guo et al. (2016) it has been shown that silica NPs generated cytotoxicity in endothelial cells from mechanisms related to oxidative stress. Although the authors postulated that, the association of antioxidant agents to the surface as functionalization agents could decrease the cytotoxic effect ensuring a safe use of these nanomaterials.

All these examples described from information gathered from the bibliography selected in time, demonstrate the clear trend that governs the behaviour of amorphous silica nanoparticles: in general, their cytotoxic effect depends on their size, their surface charge, the composition of the corona protein and its concentration. It is important to bear in mind that all these data, which, although they do not represent all the existing studies, have been carried out in vitro and on different cell types. This reinforces the conception that cell type is also a determining factor in the effects that are observed.

Cytotoxic effect of silica nanoparticles may be controlled by an exhaustive work on the design of nanoparticle surface, which may be focused on the specific biomedical application desired.

All these data were obtained from in vitro studies. However, a more realistic point of view arises when studied are performed on in vivo models.

On in vivo studies, a clearer and more definite correlation seems to be found between the size of the nanoparticles and their toxicity. Thus, in general, smaller nanoparticles seems to present higher toxicity than larger ones.

Diverse studies were performed both in vitro and in vivo between 2010 and 2015 revealed different results such as mentioned above, depending on several factors not only related to nanoparticles but also with cells.

Discrepancies between in vitro and in vivo studies may arise from designing of experiments in terms of the evaluation of improper and unrealistic doses and conditions.

A biodistribution and toxicological analysis for amorphous silica nanoparticles coated with amine groups and functionalized with fluorescent dye or radiolabelled was performed on male guinea pigs and adult male Swiss mice. A dose of 100 mg/kg intraperitoneally administered, much higher than the necessary to ensure bioavailability, was found as safe. Nanoparticles were found to biodistribute on different main organs such as liver, kidney, heart, stomach and intestine. After five days no changes were observed in organs.

A mice model was employed by Lu et al. in 2015 to evaluate the effects of size and surface modification with amine or carboxylate groups of amorphous silica nanoparticles. Smaller nanoparticles functionalized with amino groups were found as more toxic for liver. Increasing doses of micro-sized nanoparticles exerted liver toxicity independently from the surface modification agent. Both liver and spleen resulted the organs with higher concentration accumulated, showing different pathways for uptake. It was also observed the influence on toxicity associated to liver toxicity rather than size.

Different organs such as liver, heart, lung, kidney and brain were studied after acute exposure by intraperitoneally administration on mice of 50 nm-sized amorphous silica nanoparticles in dose of 0.25 mg/kg. All the organs except heart were affected by an increase of lipid peroxidation. All organs showed an intensification of superoxide dismutase and catalase activities. Tumor necrosis factor α (TNF-α) was also increased in all mentioned organs. DNA damage was also evidenced accompanied by an increase of diverse enzymes associated to oxidative stress in plasma, as well as general leucocytosis. Therefore, an acute toxicity may be adscribed to general oxidative stress.

Zhuravskii et al. (2016) conducted an in vivo study in rats to evaluate the effects of intravenous administration of small amorphous silica nanoparticles (13 nm). A dose of 7 mg/kg was administered. Hematological, biochemical and histological parameters were analysed as well as silicon content after 7, 30 and 60 days post administration. Focus was pointed on mast cells content in different organs. No alterations were determined in terms of haematological and biochemical parameters and no considerably increment in silicon was observed in different organs, except for lung only one hour after administration. It was observed a process of liver remodelling where formation of granulomas were detected after 7 days from the intravenous injection. After 30 days treatment an increase of mast cells was observed in lungs, liver and heart. Even fibrosis was detected in liver, no necrotic process occurred.

Another research performed in vivo in 2018 by Waegeneers et al. evaluated the acute effect of commercial amorphous silica nanoparticles employing a dose of 20 mg/kg. The assay was conducted on female Sprague-Dawley rats by intravenous injection. After 6 and 24 h animals were euthanized. The evaluation of silicon revealed that spleen and liver where organs containing in major quantity. Apparently nanoparticles were metabolized in liver since lower concentrations were found in spleen after the times evaluated. Macrophages were found in high amount in liver after the acute exposure. Excretion was mainly observes by urine.

In a very recent study developed by Mohammadpour et al. (2020). The chronic toxicity was evaluated during one year after intravenous administration of different silica nanoparticles in terms of size and porosity. The study was conducted on Balb-C mice of both sex. Also, human blood compatibility was evaluated in parallel.

Although no changes were observed in the hematological parameters of mice and no changes in clinical signs were observed, an exhaustive and microscopic evaluation of different organs revealed some issues related to toxicity. Clearance of nanoparticles seemed to be slow, in special in spleen and liver, so they were observed signs of inflammation in these two organs. Some observations related to calcifications in pulmonary and cardiac vessels as well as in kidney are indicative of resolved intravascular thrombosis. These accidents were mostly significant in those groups administered with larger amorphous silica nanoparticles. In relation to the haematological ex vivo study on human blood, no toxicity effects were observed related to haemolytic effects or activation of complement.

Currently, the available studies regarding the toxicity of amorphous silica nanoparticles have been carried out in vitro and in vivo in animal models. There is no accurate information about human studies that provide epidemiological data to give an idea of its biocompatibility. In addition, as it was reflected in the research cited in this chapter that there is a plenty of opportunities in the development of new works to reveal mechanisms associated to toxicity of amorphous silica nanoparticles. This is considering the versatility they present as nanomedicines platforms for the therapeutics of several diseases aiming to improve current treatments in general.

There is not enough information in relation to the effect of coating to improve biocompatibility. This point seems to be an important window for future research.

Regarding the applications on bone, in the next chapter a revision in terms of bioactivity is bone will be presented. And it is worth to mention that all these queries here discussed associated to toxicology may be avoided if nanoparticles may be used during surgery procedures, in special in bone cancer disease and pathologies associated with implants infections.

References

J.F. Affonso de Oliveira, F.R. Scheffer, R.F. Landis, E. Teixeira Neto, V.M. Rotello, M.B. Cardoso, Dual functionalization of nanoparticles for generating corona-free and noncytotoxic silica nanoparticles. ACS Appl. Mater. Interfaces **10**(49), 41917–41923 (2018)

W.J. Akers, M.Y. Berezin, H. Lee, K. Guo, A. Almutairi, J.M. Fréchet, G.M. Fischer, E. Daltrozzo, S. Achilefu, Biological applications of fluorescence lifetime imaging beyond microscopy, in *Reporters, Markers, Dyes, Nanoparticles, and Molecular Probes for Biomedical Applications II*, vol. 7576 (International Society for Optics and Photonics, 2010), p. 757612

R.P. Bagwe, L.R. Hilliard, W. Tan, Surface modification of silica nanoparticles to reduce aggregation and nonspecific binding. Langmuir **22**(9), 4357–4362 (2006)

G. R. Beck Jr, S.W. Ha, C.E. Camalier, M. Yamaguchi, Y. Li, J.K. Lee, M.N. Weitzmann, Bioactive silica-based nanoparticles stimulate bone-forming osteoblasts, suppress bone-resorbing osteoclasts, and enhance bone mineral density in vivo. Nanomed. Nanotechnol. Biol. Med. **8**(6), 793–803 (2012)

C.D. Brites, V.T. Freitas, R.A. Ferreira, A. Millán, F. Palacio, L.D. Carlos, Metal-free highly luminescent silica nanoparticles. Langmuir **28**(21), 8190–8196 (2012)

T. Cedervall, I. Lynch, M. Foy, T. Berggad, S. Donnelly, G. Cagney, S. Linse, K. Dawson, Detailed identification of plasma proteins adsorbed on copolymer nanoparticles. Angew. Chem. Int. Ed. **46**, 5754–5756 (2007)

M.H. Chan, H.M. Lin, Preparation and identification of multifunctional mesoporous silica nanoparticles for in vitro and in vivo dual-mode imaging, theranostics, and targeted tracking. Biomaterials **46**, 149–158 (2015)

S. Chandra, G. Beaune, N. Shirahata, F.M. Winnik, A one-pot synthesis of water soluble highly fluorescent silica nanoparticles. J. Mater. Chem. B **5**(7), 1363–1370 (2017)

A.B. Chinen, C.M. Guan, J.R. Ferrer, S.N. Barnaby, T.J. Merkel, C.A. Mirkin, Nanoparticle probes for the detection of cancer biomarkers, cells, and tissues by fluorescence. Chem. Rev. **115**(19), 10530–10574 (2015)

Y.E. Choi, J.W. Kwak, J.W. Park, Nanotechnology for early cancer detection. Sensors **10**(1), 428–455 (2010). Berezin, 2010

J.J. Corbalan, C. Medina, A. Jacoby, T. Malinski, M.W. Radomski, Amorphous silica nanoparticles trigger nitric oxide/peroxynitrite imbalance in human endothelial cells: inflammatory and cytotoxic effects. Int. J. Nanomed. **6**, 2821 (2011)

J.M. Garcia, M.A. Mondragon, C.S. Téllez, A. Campero, V.M. Castano, Blue emission in tetraethoxysilane and silica gels. Mater. Chem. Phys. **41**(1), 15–17 (1995)

M. Gisbert-Garzarán, M. Manzano, M. Vallet-Regí, Mesoporous silica nanoparticles for the treatment of complex bone diseases: bone cancer, bone infection and osteoporosis. Pharmaceutics **12**(1), 83 (2020)

W.H. Green, K.P. Le, J. Grey, T.T. Au, M.J. Sailor. White phosphors from a silicate-carboxylate sol-gel precursor that lack metal activator ions. Science **276**(5320), 1826–1828 (1997)

C. Guo, M. Yang, L. Jing, J. Wang, Y. Yu, Y. Li, …, Z. Sun, Amorphous silica nanoparticles trigger vascular endothelial cell injury through apoptosis and autophagy via reactive oxygen species-mediated MAPK/Bcl-2 and PI3K/Akt/mTOR signaling. Int. J. Nanomedicine **11**, 5257 (2016)

N.J. Halas, Nanoscience under glass: the versatile chemistry of silica nanostructures. ACS Nano **2**(2), 179–183 (2008)

A.M. Jakob, T.A. Schmedake, A novel approach to monodisperse, luminescent silica spheres. Chem. Mater. **18**(14), 3173–3175 (2006)

L. Jiang, Y. Yu, Y. Li, Y. Yu, J. Duan, Y. Zou, Q. Li, Z. Sun, Oxidative damage and energy metabolism disorder contribute to the hemolytic effect of amorphous silica nanoparticles. Nanoscale Res. Lett. **11**(1), 57 (2016)

A. Kurtz-Chalot, C. Villiers, J. Pourchez, D. Boudard, M. Martini, P.N. Marche, M. Cottier, V. Forest, Impact of silica nanoparticle surface chemistry on protein corona formation and consequential interactions with biological cells. Mater. Sci. Eng. C **75**, 16–24 (2017)

A. Lankoff, M. Arabski, A. Wegierek-Ciuk, M. Kruszewski, H. Lisowska, A. Banasik-Nowak, K. Rozga-Wijas,. M. Wojewodzka, S. Slomkowski, Effect of surface modification of silica nanoparticles on toxicity and cellular uptake by human peripheral blood lymphocytes in vitro. Nanotoxicology, **7**(3), 235–250 (2012)

R. Leibe, I.L. Hsiao, S. Fritsch-Decker, U. Kielmeier, A.M. Wagbo, B. Voss, A. Schmidt, S.D. Hessman, A. Duschl, G.J. Oostingh, S. Diabaté, The protein corona suppresses the cytotoxic and pro-inflammatory response in lung epithelial cells and macrophages upon exposure to nanosilica. Arch. Toxicol. **93**(4), 871–885 (2019)

A. Lesniak, F. Fenaroli, M.P. Monopoli, C. Åberg, K.A. Dawson, A. Salvati, Effects of the presence or absence of a protein corona on silica nanoparticle uptake and impact on cells. ACS Nano **6**(7), 5845–5857 (2012)

X. Lu, C. Ji, T. Jin, X. Fan, The effects of size and surface modification of amorphous silica particles on biodistribution and liver metabolism in mice. Nanotechnology **26**(17), 175101 (2015)

I. Lynch, K.A. Dawson, Protein-nanoparticle interactions. Nano Today **3**(1–2), 40–47 (2008)

K.R. Martin, The chemistry of silica and its potential health benefits. J. Nutr. Health Aging **11**(2), 94 (2007)

A. Mendoza, J.A. Torres-Hernandez, J.G. Ault, J.H. Pedersen-Lane, D. Gao, D.A. Lawrence, Silica nanoparticles induce oxidative stress and inflammation of human peripheral blood mononuclear cells. Cell Stress Chaperones **19**(6), 777–790 (2014)

R. Mohammadpour, D.L. Cheney, J.W. Grunberger, M. Yazdimamaghani, J. Jedrzkiewicz, K.J. Isaacson, M.A. Dobrovolskaia, H. Ghandehari, One-year chronic toxicity evaluation of single dose intravenously administered silica nanoparticles in mice and their Ex vivo human hemocompatibility. J. Control. Release (2020)

S. Murugadoss, D. Lison, L. Godderis, S. Van Den Brule, J. Mast, F. Brassinne, N. Sebaihi, P.H. Hoet, Toxicology of silica nanoparticles: an update. Arch. Toxicol. **91**(9), 2967–3010 (2017)

D. Napierska, L.C. Thomassen, V. Rabolli, D. Lison, L. Gonzalez, M. Kirsch-Volders, J.A. Martens, P.H. Hoet, Size-dependent cytotoxicity of monodisperse silica nanoparticles in human endothelial cells. Small **5**(7), 846–853 (2009)

A. Nemmar, S. Beegam, P. Yuvaraju, J. Yasin, A. Shahin, B.H. Ali, Interaction of amorphous silica nanoparticles with erythrocytes in vitro: role of oxidative stress. Cell. Physiol. Biochem. **34**(2), 255–265 (2014)

A. Nemmar, P. Yuvaraju, S. Beegam, J. Yasin, R. Al Dhaheri, M.A. Fahim, B.H. Ali, In vitro platelet aggregation and oxidative stress caused by amorphous silica nanoparticles. Int. J. Physiol. Pathophysiol. Pharmacol. **7**(1), 27 (2015)

A. Nemmar, P. Yuvaraju, S. Beegam, J. Yasin, E.E. Kazzam, B.H. Ali, Oxidative stress, inflammation, and DNA damage in multiple organs of mice acutely exposed to amorphous silica nanoparticles. Int. J. Nanomed. **11**, 919 (2016)

I.A. Rahman, P. Vejayakumaran, C.S. Sipaut, J. Ismail, M.A. Bakar, R. Adnan, C.K. Chee, An optimized sol–gel synthesis of stable primary equivalent silica particles. Colloids Surf. A **294**(1–3), 102–110 (2007)

K.S. Rao, K. El-Hami, T. Kodaki, K. Matsushige, K. Makino, A novel method for synthesis of silica nanoparticles. J. Colloid Interface Sci. **289**(1), 125–131 (2005)

J. Saikia, M. Yazdimamaghani, S.P. Hadipour Moghaddam, H. Ghandehari, Differential protein adsorption and cellular uptake of silica nanoparticles based on size and porosity. ACS Appl. Mater. Interfaces **8**(50), 34820–34832 (2016)

H. Shinto, T. Fukasawa, K. Yoshisue, M. Tezuka, M. Orita, Cell membrane disruption induced by amorphous silica nanoparticles in erythrocytes, lymphocytes, malignant melanocytes, and macrophages. Adv. Powder Technol. **25**(6), 1872–1881 (2014)

H. Shinto, T. Fukasawa, K. Yoshisue, N. Tsukamoto, S. Aso, Y. Hirohashi, H. Seto, Effect of interfacial serum proteins on the cell membrane disruption induced by amorphous silica nanoparticles in erythrocytes, lymphocytes, malignant melanocytes, and macrophages. Colloids Surf. B **181**, 270–277 (2019)

A. Sikora, A.G. Shard, C. Minelli, Size and ζ-potential measurement of silica nanoparticles in serum using tunable resistive pulse sensing. Langmuir **32**(9), 2216–2224 (2016)

W. Stöber, A. Fink, E. Bohn, Controlled growth of monodisperse silica spheres in the micron size range. J. Colloid Interface Sci. **26**(1), 62–69 (1968)

B.I. Tamba, A. Dondas, M. Leon, A.N. Neagu, G. Dodi, C. Stefanescu, A. Tijani, Silica nanoparticles: preparation, characterization and in vitro/in vivo biodistribution studies. Eur. J. Pharm. Sci. **71**, 46–55 (2015)

D.A. Tomalia, Birth of a new macromolecular architecture: dendrimers as quantized building blocks for nanoscale synthetic polymer chemistry. Prog. Polym. Sci. **30**(3–4), 294–324 (2005)

A. Van Blaaderen, A. Vrij, Synthesis and characterization of colloidal dispersions of fluorescent, monodisperse silica spheres. Langmuir **8**(12), 2921–2931 (1992)

N. Waegeneers, A. Brasseur, E. Van Doren, S. Van der Heyden, P.J. Serreyn, L. Pussemier, J. Mast, Y.J. Schneider, A. Ruttens, S. Roels, Short-term biodistribution and clearance of intravenously administered silica nanoparticles. Toxicol. Rep. **5**, 632–638 (2018)

X. Wang, A.R. Morales, T. Urakami, L. Zhang, M.V. Bondar, M. Komatsu, K.D. Belfield, Folate receptor-targeted aggregation-enhanced near-IR emitting silica nanoprobe for one-photon in vivo and two-photon ex vivo fluorescence bioimaging. Bioconjugate Chem. **22**(7), 1438–1450 (2011)

S.H. Wu, C.Y. Mou, H.P. Lin, Synthesis of mesoporous silica nanoparticles. Chem. Soc. Rev. **42**(9), 3862–3875 (2013)

H. Xu, F. Yan, E.E. Monson, R. Kopelman, Room-temperature preparation and characterization of poly (ethylene glycol)-coated silica nanoparticles for biomedical applications. J. Biomed. Mater. Res. Part A: Official J. Soc. Biomater. Jpn. Soc. Biomater. Aust. Soc. Biomater. Korean Soc. Biomater. **66**(4), 870–879 (2002)

L. Xue, B. Li, Q. Fei, G. Feng, Y. Huan, Z. Shi, Carboxylate-modified squaraine dye doped silica fluorescent pH nanosensors. Nanotechnology **21**(21), 215502 (2010)

J. Yao, M. Yang, Y. Duan, Chemistry, biology, and medicine of fluorescent nanomaterials and related systems: new insights into biosensing, bioimaging, genomics, diagnostics, and therapy. Chem. Rev. **114**(12), 6130–6178 (2014)

B.E. Yoldas, Photoluminescence in chemically polymerized SiO_2 and Al_2O_3–SiO_2 systems. Blue emission in tetraethoxysilane and silica gels White phosphors from a silicate-carboxylate sol-gel precursor that lack metal activator ions. J. Mater. Res. **5**(6), 1157–1158 (1990)

T. Yu, A. Malugin, H. Ghandehari, Impact of silica nanoparticle design on cellular toxicity and hemolytic activity. ACS Nano. **5**(7), 5717–5728 (2011)

S. Zhuravskii, G. Yukina, O. Kulikova, A. Panevin, V. Tomson, D. Korolev, M. Galagudza, Mast cell accumulation precedes tissue fibrosis induced by intravenously administered amorphous silica nanoparticles. Toxicol. Mech. Methods **26**(4), 260–269 (2016)

Chapter 6
Bioactive Properties of Amorphous Silica Nanoparticles in Bone

This chapter is intended to show how amorphous silica nanoparticles have bioactive properties on bone tissue.

In the design of nanomaterials for biomedical applications, biocompatible and versatile nanomaterials are sought to be chemically modified to achieve the properties with the desired functionality. But when we talk about amorphous silica nanoparticles we have the plus that bioactively they can help bone tissue in addition to fulfilling their function as nanoplatforms for drug delivery or issues associated with tissue engineering.

For a pleasant discussion, some of the most relevant works since 2011 have been selected to provide evidence at the mechanical and molecular level of how nanoparticles exert their effects on bone cells.

In the first chapters of the book, a general overview was presented about the functions of each of the cells that make up bone tissue. This information is very useful in this chapter to understand the mechanisms that trigger amorphous silica nanoparticles related to the stimulation of bone regeneration.

From the physicochemical point of view of nanoparticles, it will be mentioned how certain properties such as size and surface charge influence on bioactivity. These issues are highly relevant to the design of optimal nanomaterials with specific properties.

Nabeshi et al. (2011) worked on the effect of amorphous silica nanoparticles of different sizes (70, 100 and 1000 nm) on a murine macrophage RAW264.7 cell line. Observation points were based on internalization, intracellular localization sites, cytotoxity, and induction of differentiation of this macrophage cell line in osteoclasts. Smaller 70 nm nanoparticles resulted as more cytotoxic than larger ones at concentrations of 30 µg/mL while this effect was lower at 10 µg/mL. In this work it was observed that the internalization of the nanoparticles is greater with decreasing their size. Even those of smaller size were observed in the cell nucleus. The effect on osteoclast differentiation in RAW264.7 cells was evaluated adding in culture media

M. Agotegaray, *Silica-Based Nanotechnology for Bone Disease Treatment*,
SpringerBriefs in Applied Sciences and Technology,
https://doi.org/10.1007/978-3-030-64130-6_6

RANKL, considering it is a key molecule in osteoclast differentiation and TRAP-positive multinucleated cells were considered as indicators of differentiation. Authors observed no differentiation in presence of neither of the nanoparticles studied, except for 70 nm-sized. This work provides key data regarding with the fact that amorphous silica nanoparticles do not induce differentiation to osteoclasts. This is very important for their potential employment in biomedicine.

The importance of silicon and its fundamental role in bone metabolism have been mentioned earlier in this book. Reason why it is a very important element in the diet. However, until very recently, there was little knowledge about the role of silica nanoparticles in this regard.

In 2012, Beck Jr. et al. obtained significant evidence of the beneficiating role of 50 nm-sized amorphous silica nanoparticles for bone metabolism. In their work, it was demonstrated an inhibitory action on osteoclasts. This effect is mediated by the capacity to antagonize the activation of NF-κB which is a factor involved in resorption exerted by osteoclasts. In addition, it was shown that silica nanoparticles induced osteoblast differentiation and an increased in bone mineral density.

Working with biocompatible coatings on the surface of silica nanoparticles is also a good alternative for developing bioactive silica-based nanomaterials for bone.

And the evaluation of its functionality is extremely important. In this regard, Amorim et al. (2014) coated amorphous silica nanoparticles with two molecules with relevance for bone: hyaluronic acid and poly-L-lysine. Hyaluronic acid is an important structural element of the bone matrix. Its role is related to the induction of migration, anchorage and proliferation of bone cells through the activation of specific intracellular mechanisms after interaction with membrane receptors, such as CD44. Meanwhile, poly-L-lysine is widely used in cell culture to promote adhesion. The combination of both for surface coating of silica nanoparticles lead to particle hydrodynamic diameter of near 570 nm. These nanoparticles were assayed on culture of human bone marrow stem cells to evaluate both cytotoxicity and osteogenic differentiation. After 21 days of exposure to different concentrations ranging from 12.5 to 50 mg/mL it was demonstrated that after treatment cells overexpressed diverse factors involved in osteogenesis such as Runx-2, alkaline phosphatase ALP, sialoprotein, osteopontin and collagen type I in special for 12.5 and 25 mg/ml treatments. No cytotoxic effect was observed revealing a trend to induce differentiation of the cell studied.

An in vivo research on young mice conducted by Ha et al. in 2014 revealed that silica nanoparticles induced differentiation of osteoblasts and promoted mineralization. This observation was associated to nanoparticles uptake by an endocytic pathway mediated by caveolae. Activation of kinase ERK1/2 would lead to autophagosome formation which may be associated with differentiation and mineralization processes.

Physicochemical properties of nanoparticles such as surface charge are a key factor in determining bioactivity of nanoparticles. In this regards, it is necessary to impart specific features o nanoparticles surface to ensure internalization into bone cells. Shahabi et al. (2015) designed different types of silica nanoparticles, surface functionalized with different groups to impart differences in surface charge and thus evaluate the influence on uptake by human osteoblasts. Amino and sulphonate groups lead to positive and negative surface charged nanoparticles, respectively. It was observed that serum proteins interacted with nanoparticles preventing from aggregation. Sulphonate groups induced a better uptake when serum proteins were present. This is important because once again it is worth to mention that considering the presence of protein corona is a key factor for evaluation of nanoparticles intended for biomedical applications. In absence of serum proteins, amino-functionalized nanoparticles rendered better uptake. But this is strictly concerning nanoparticle point of view. The influence of medium and the characteristic of cell are also determinant parameters in biological media.

Promotion of bone formation was observed in vivo in mice in a study performed in 2015 by Weitzmann et al. This model resulted as very interesting because it involved aged mice and it was observed that 50 nm-sized silica nanoparticles induced an increase in bone mineral density, volume and they were also observed an increment in bone formation markers.

Continuing with the bioactive effect of silica nanoparticles on bone, also Filipowska et al. in 2018 showed that the presence of silica nanoparticles induces the expression of genes and factors related to osteogenesis. In this case they produced nano-hydrogels sized in 240 and 450 nm, based on silica nanoparticles hybridized with collagen and chitosan. The analysis on marrow-derived mesenchymal stromal cells revealed all tested materials not only as biocompatible, but also as inductors of the expression of RUNX-2, osteocalcin and VEGF. In addition, after 20 days not only differentiation was observed but also a mineralization process was evidenced.

In the spirit of deepening the effect of the physicochemical properties of amorphous silica nanoparticles on osteogenesis, Ha et al. (2018) prepared diverse formulations varying not only the size, but also the surface functionalization to observe their influence. From 50 to 450 nm sized silica nanoparticles and with different surface groups such as hydroxyl, carboxylate and amine. In this work, authors revealed that osteoblast differentiation is mainly influenced by both composition and size, being 50 nm sized-nanoparticles those with better performance. On the other hand, the inhibition of osteoclastogenesis was herein related to the surface charge of nanoparticles, being negative charge necessary to this end. The mechanism evidenced for this effect was associated to the inhibition of transcriptional factor Nfatc1 which blank gene is RANK by silica nanoparticles.

From the bibliographic analysis, a clear trend can be evidenced that affirms the bioactive properties of amorphous silica nanoparticles in bone.

These properties are related to the stimulation of osteogenic activity and the inhibition of osteoclastic activity by different mechanisms. Almost all related to gene expression. In turn, it is extremely important to keep in mind that the physicochemical properties of nanoparticles drastically influence their effect. In this way, it is evident not only the great potential of these nanoparticles as nanotechnological platforms intended to improve current treatments, but also as therapeutic agents in themselves for bone.

References

S. Amorim, A. Martins, N.M. Neves, R.L. Reis, R.A. Pires, Hyaluronic acid/poly-L-lysine bilayered silica nanoparticles enhance the osteogenic differentiation of human mesenchymal stem cells. J. Mater. Chem. B **2**(40), 6939–6946 (2014)

G.R. Beck Jr., S.W. Ha, C.E. Camalier, M. Yamaguchi, Y. Li, J.K. Lee, M.N. Weitzmann, Bioactive silica-based nanoparticles stimulate bone-forming osteoblasts, suppress bone-resorbing osteoclasts, and enhance bone mineral density in vivo. Nanomed. Nanotechnol. Biol. Med. **8**(6), 793–803 (2012)

J. Filipowska, J. Lewandowska-Łańcucka, A. Gilarska, Ł. Niedźwiedzki, M. Nowakowska, In vitro osteogenic potential of collagen/chitosan-based hydrogels-silica particles hybrids in human bone marrow-derived mesenchymal stromal cell cultures. Int. J. Biol. Macromol. **113**, 692–700 (2018)

S.W. Ha, M.N. Weitzmann, G.R. Beck Jr., Bioactive silica nanoparticles promote osteoblast differentiation through stimulation of autophagy and direct association with LC3 and p62. ACS Nano **8**(6), 5898–5910 (2014)

S.W. Ha, M. Viggeswarapu, M.M. Habib, G.R. Beck Jr., Bioactive effects of silica nanoparticles on bone cells are size, surface, and composition dependent. Acta Biomater. **82**, 184–196 (2018)

H. Nabeshi, T. Yoshikawa, T. Akase, T. Yoshida, S. Tochigi, T. Hirai, M. Uji, K.I. Ichihashi, T. Yamashita, K. Higashisaka, Y. Morishita, Effect of amorphous silica nanoparticles on in vitro RANKL-induced osteoclast differentiation in murine macrophages. Nanoscale Res. Lett. **6**(1), 464 (2011)

S. Shahabi, L. Treccani, R. Dringen, K. Rezwan, Modulation of silica nanoparticle uptake into human osteoblast cells by variation of the ratio of amino and sulfonate surface groups: effects of serum. ACS Appl. Mater. Interfaces. **7**(25), 13821–13833 (2015)

M.N. Weitzmann, S.W. Ha, T. Vikulina, S. Roser-Page, J.K. Lee, G.R. Beck Jr., Bioactive silica nanoparticles reverse age-associated bone loss in mice. Nanomed. Nanotechnol. Biol. Med. **11**(4), 959–967 (2015)

Chapter 7
Amorphous Silica Nanoparticles as Bone Therapeutics

The use of nanotechnologies aims to fill the gaps that currently exist in the treatment of various bone pathologies, especially osteoporosis, bone cancer and infectious diseases.

Many attempts are still under investigation in relation to the development of nanomaterials that contribute to this problem. In this way, a lot of work has been done with materials based on hydroxyapatite, polymers, ceramics, mesoporous silica nanomaterials, xerogels that tend to try to solve this problem.

But delving into the functionality of the amorphous silica nanoparticles in this context, it is observed that there are not many developments in this field even taking into account the magnificent properties related to their bioactive effect at bone level.

For this reason, in this chapter, the main clinical problems that need to be solved today will be described and the contributions that exist in relation to amorphous silica nanomaterials in these aspects will be discussed.

One of the most clinical problem worldwide is related to replacement of bone either due to damage or of lost bone. Fractures or tumors are the main causes for these troubles.

In addition to the high costs of the treatments associated with this problem, the morbidity of the patients and the low quality of life, in general, the post-surgical problems associated with infections occur.

In this regards, tissue engineering comes to provide solutions to this entire problem.

The functionality of synthetic bone grafts is based on a complex process in which a scaffold is necessary to promote and induce cell growth with participation of bioactive factors. In this process it necessary the formation of bone matrix and the role of osteoprogenitor cells is key for a successful progress in remodelling and healing. Nanomaterials arises as promising helpers for these issues. And amorphous silica nanoparticles play a key role in this regards. These nanoparticles were employed in different research works to improve the quality of different materials for bone regeneration.

© The Author(s), under exclusive license to Springer Nature Switzerland AG 2020
M. Agotegaray, *Silica-Based Nanotechnology for Bone Disease Treatment*,
SpringerBriefs in Applied Sciences and Technology,
https://doi.org/10.1007/978-3-030-64130-6_7

Hesaraki et al. in 2011 studied the influence of nano-amorphous silica by producing a composite with hydroxyapatite. The obtained material was a slurry that presented improved properties in comparison to hydroxyapatite both in relation to mechanical properties as well as to biological effects. Precipitation of apatite nanocrystals on the composite was successful and proliferation of osteoblasts was also observed presenting higher phosphatase alkaline activity.

Mohammadi et al. obtained a calcium phosphate cement added with amorphous silica nanoparticles in 2014 with optimum properties in terms of compressive strength imparted by the addition of the nanoparticles.

Injectable hydrogels also represent desirable materials for bone regeneration. Chitosan and collagen are biopolymers of choice for construction of this type of materials. Amorphous silica nanoparticles presented a positive effect on these materials.

Thus, Lewandowska et al. (2015a) dispersed these nanoparticles in different hydrogels matrixes of chitosan and collagen as well as in combination of both. Cell viability was observed as improved by the presence of nanosilica. In addition, the mineralization process was proven to be incremented by their presence.

As mentioned earlier in this book, silicon plays a fundamental role in the entire process of cellular regeneration through the stimulation of intracellular mechanisms. Moreover, it has been proven that amorphous silica materials have the ability to promote the regeneration process precisely from their silicon-based composition. Silicon ions were released in vitro after 24 h as observed in vitro and this was associated to the increase of factors related to osteogenesis since activation transcription 4, collagen and alkaline phosphatase were incremented. In addition, Ca-P precipitation was also evidenced after 85 days of study on human periosteum cells.

In relation to the importance of silicon on the increasing effects regarding tissue regeneration, these effects were imparted by amorphous silica nanoparticles. In vitro release of orthosilicic acid [$Si(OH)_4$] was observed by Quignard et al. in 2017, from 10 nm-sized silica nanoparticles surface functionalized with different groups. This behaviour induced increased viability, proliferation and migration of fibroblasts in a wound healing model. In special, functionalization with amine groups resulted as more effective. This effect was ascribed to nanoparticles uptake by cells, which would facilitate dissolution.

Reactive oxygen species were found to be exacerbated in traumatic fractures. Osteoblast differentiation is very difficult under these conditions. So, other pathway to induce osteogenesis stimulation could be related to the increase in the activity of natural antioxidants as such is the case of superoxide dismutase. Thus, a novel healing biomaterial containing amorphous silicon oxynitride was proposed as a possible and promising option to deal with oxidative stress condition. The sustained release of Si^{+4} is responsible for the healing properties of this material as demonstrated by Ilyas et al. in a research work developed in 2016. In addition, an increase of SOD activity was evidenced as responsible for increasing bone regeneration and mineral deposition.

Elastomeric behaviour to ensure a proper medium for stimulating cellular activity is a property necessary for bone healing materials. In this regards it has been observed that the in situ formation of amorphous silica nanoparticles poly(citrate-siloxane)hybrid elastomers improves therapeutic properties of these materials. The dispersion of the nanoparticles in the matrix of the scaffold increased biocompatibility on osteoblasts and imparted photoluminscent properties, which are very useful for therapeutics and evaluation of progression in the treatment.

In a very recent research work, Ballarre et al. (2020) worked on titanium orthopaedic implants in order to improve their characteristics as bone implant material. By spray and electrophorectic deposition, of chitosan-gelatin silica containing gentamicin. Coating based on amorphous silica nanoparticles was responsible for an increased cell viability while gentamicin reduced bacteria growing.

Based on the above, it can be seen that amorphous silica nanoparticles play an indisputable role in the process of bone cell differentiation and osteogenesis. They even have the property of improving the physicochemical characteristics of materials that are already known to have a suitable effect to be implemented in bone prostheses or in fillers.

The important to note here is that there is a niche that needs to be explored regarding the development of simple nanomaterials based on amorphous silica for the treatment of bone pathologies.

There are not so many works that have been dedicated to this topic. When exploring the bibliography, mesoporous silica does turn out to be a very studied nanomaterial. However, amorphous silica presents itself as a very promising option. Thus, by focusing scientific work on this simple and noble material, many avenues are opened to provide necessary, low-cost and promising therapeutic solutions.

References

J. Ballarre, T. Aydemir, L. Liverani, J.A. Roether, W.H. Goldmann, A.R. Boccaccini, Versatile bioactive and antibacterial coating system based on silica, gentamicin, and chitosan: improving early stage performance of titanium implants. Surf. Coat. Technol. **381**, 125138 (2020)

S. Hesaraki, H. Nazarian, M. Alizadeh, Multi-phase biocomposite material in-situ fabricated by using hydroxyapatite and amorphous nanosilica. Int. J. Mater. Res. **102**(5), 494–503 (2011)

A. Ilyas, T. Odatsu, A. Shah, F. Monte, H.K. Kim, P. Kramer, P.B. Aswath, V.G. Varanasi, Amorphous silica: a new antioxidant role for rapid critical-sized bone defect healing. Adv. Healthc. Mater. **5**(17), 2199–2213 (2016)

J. Lewandowska-Łańcucka, S. Fiejdasz, Ł. Rodzik, M. Kozieł, M. Nowakowska, Bioactive hydrogel-nanosilica hybrid materials: a potential injectable scaffold for bone tissue engineering. Biomed. Mater. **10**(1), 015020 (2015a)

J. Lewandowska-Łańcucka, S. Fiejdasz, Ł. Rodzik, A. Łatkiewicz, M. Nowakowska, Novel hybrid materials for preparation of bone tissue engineering scaffolds. J. Mater. Sci. Mater. Med. **26**(9), 231 (2015b)

Y. Li, Y. Guo, J. Ge, P.X. Ma, B. Lei, In situ silica nanoparticles-reinforced biodegradable poly (citrate-siloxane) hybrid elastomers with multifunctional properties for simultaneous bioimaging and bone tissue regeneration. Appl. Mater. Today **10**, 153–163 (2018)

M. Mohammadi, S. Hesaraki, M. Hafezi-Ardakani, Investigation of biocompatible nanosized materials for development of strong calcium phosphate bone cement: comparison of nano-titania, nano-silicon carbide and amorphous nano-silica. Ceram. Int. **40**(6), 8377–8387 (2014)

T. Odatsu, T. Azimaie, M.F. Velten, M. Vu, M.B. Lyles, H.K. Kim, P.B. Aswath, V.G. Varanasi, Human periosteum cell osteogenic differentiation enhanced by ionic silicon release from porous amorphous silica fibrous scaffolds. J. Biomed. Mater. Res. Part A **103**(8), 2797–2806 (2015)

S. Quignard, T. Coradin, J.J. Powell, R. Jugdaohsingh, Silica nanoparticles as sources of silicic acid favoring wound healing in vitro. Colloids Surf. B **155**, 530–537 (2017)

Chapter 8
Perspectives

The spirit of this book lies in considering bone pathologies as a scourge that affects many people in the world. And not only to those patients who suffer them, but to doctors and scientific researchers who, day by day, dedicate their time and effort to improve the existing conditions for treatment.

Actually, the most important pathologies that afflict the bone system are painful, chronic, and difficult to access and treat. Osteoporosis, cancer and infections once diagnosed do not have a good prognosis. Added to that there are not many tools to alleviate the symptoms and the difficult living conditions that they bring.

For this reason, the contribution of this book aims to highlight one more alternative aiming to find tools that help to improve the management of bone diseases. Thus, the silica nanoparticles are presented as promising and friendly for this purpose.

Their study, the exploitation of their versatility in terms of drug loading and the use of the bioactive properties in bone are a niche to going on deepen in scientific research that tries to discover new ways to treat bone pathologies. Surface lability to functionalize with bone therapeutics, antibiotics and biomolecules in general arises as potential point to investigate.

In addition to this, the use of canine animal models that unfortunately suffer so much from osteosarcoma and all the tools that today allow to model tumors would be an ideal complement to fulfil this objective.

Fortunately, in our laboratory we are dedicated to this work. And I hope that with this material, intended to allow everyone to discover the potential of amorphous silica that has not yet been deeply explored, we can contribute in the short term from science so that bone pathologies do not represent a scourge. And in this way helping those who suffer in this regard to find relief.

M. Agotegaray, *Silica-Based Nanotechnology for Bone Disease Treatment*,
SpringerBriefs in Applied Sciences and Technology,
https://doi.org/10.1007/978-3-030-64130-6_8

Reference

T.M. Fan, R.D. Roberts, M.M. Lizardo, Understanding and modeling metastasis biology to improve therapeutic strategies for combating osteosarcoma progression. Front. Oncol. **10** (2020)